DO THESE STA
GO UP OR DOWN?

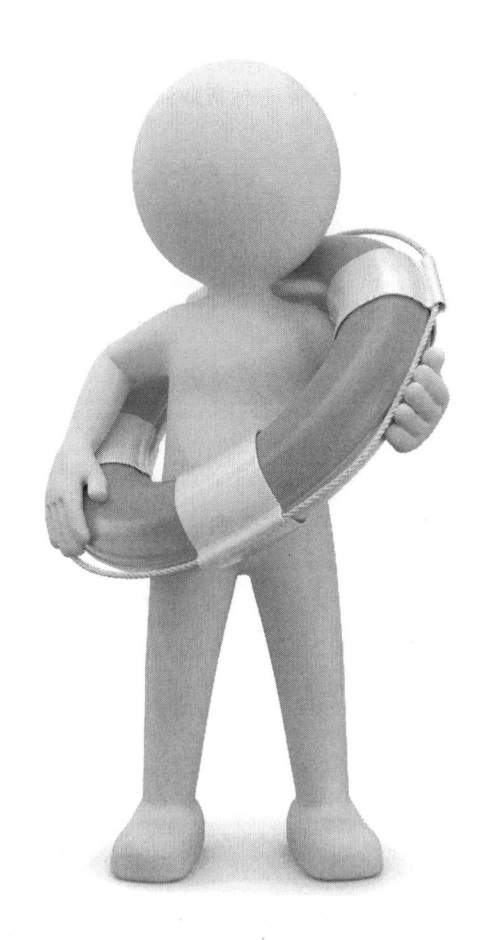

ROBIN BOLTMAN

1st Printing July 2007
2nd Revised Edition May 2020

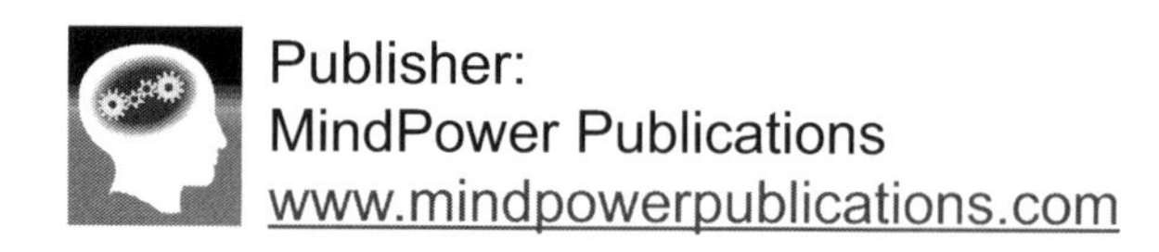
Publisher:
MindPower Publications
www.mindpowerpublications.com

Email: robinboltman@gmail.com

ISBN: 9798677441875

Cover photograph of the Oceanos by the kind consent of the copyright holder through her agent Debbie McLean.

Achille Lauro photographs by Brandon Bromfield and are available from his studio collection.
Photographs supplied by friends, passengers,Ships photographers and the SAAF

Cover design by Wolfgang Riebe
Front Cover - Oceanos
Back Cover - Robin Boltman

DEDICATION

To my Mom, Joan Francis Boltman, who has worked hard all her life, raising children, grand-children and baby sitting into her twilight years. She is still on duty today. To my late Dad - Leslie Boltman, who was probably the hardest worker I have ever known. In rain, wind and sunshine, he would walk to work from the top of University Est. to the docks and often return in the dark. If you stopped him anywhere and asked after his favourite song, no matter what sort of day he had had being a ships plumber, he would reply, "What a Wonderful World."

I love you and thank you both.

Robin

CHAPTERS

Contents **Page**

FAST FOREWARD
by Barry Hilton

I can hardly talk English... so imagine my surprise when I was approached to write a foreword for South Africa's finest magician. It's a great privilege and honour to introduce Robin Boltman's book highlighting his exploits around the world. He's an immensely talented entertainer and a really cool guy. Robin and I went to the same school in Cape Town, but at different times. We met a couple of years later when he was doing his regular comedy and magic shows with the resident band at the Hout Bay Hotel. By that time everybody knew *Mr. Magish.*

I remember him though as the Nedbank teller at the Salt River branch. We used to tell each other, "One day we'll be famous and whoever gets there first will get the other one on his TV Show!" (Makes you think doesn't it?) Robin has always been renowned for his generosity, kind nature and positive outlook - he is an inspiration to everyone in the entertainment industry. He is also the only magician I know to make two ships disappear - David Copperfield eat your heart out! The only thing I hold against him is that he took his guitar away from me as I couldn't afford to buy it. So off he went on the Achille Lauro and now our precious guitar is at the bottom of the ocean - instead of in my studio. I have been a professional for 25 years and continue to hear only great things about Robin's brilliant magic and comedy and his ability to keep audiences mystified by his incredible talents. This book will undoubtedly take you on an unforgettable adventurous journey - I hope you enjoy it as much as I enjoy being his *cousin* and pal.

The Cousin, Barry Hilton

VERY FOREWORD by Arnold Geerdts

I was extremely worried when Boltman asked me to write something for his book. A book from Boltman's pen is easy to preface. Surely this is the type of man all men hope of being? I certainly and honestly know that I do - and I always have. A man who jumps to the rescue when all else has failed. When a dire situation has arisen and there is only one thing to be: a hero. Such is Boltman. Twice!

Surely every little boy is born dreaming of what he has done... heck even grown men dream of this. A ship full of people: a dark stormy night: Radio crackling: The Captain absconds: Dashing blond hero takes control, runs around organising, and saves the day. We all long for that scenario in some sense. But - if and when it arises - how many will actually rise to this extreme challenge, risking your own life to save others? I don't truly know if I have the fibre. I live on hope, but Boltman has proved it... twice!

But that my dear friends and lucky owner of this amazing bit of literature, is, as they say in the classics, not all. No Sir! My friend Boltman is one of those people who you could end up meeting at a party in Durbanville in Cape Town, or a pub in Durban, not knowing him from the proverbial bar of soap. You would however have met him at the bar.

But on reflection, when you stagger away at two in the morning... being supported by the personage who got you in this state in the first place (Boltman), you'd think you knew him all your life! Yep, he is one of the few people who could actually speak to you in a crowded room, enquire about your health, the family's and the cat's and make you feel like you are truly the only one in the room.

But still - wait, there's more: He is undoubtedly the funniest man I know. This guy was born funny.

I first met him on a cruise to Mozambique, on the ill-fated Achille Lauro, in 1991. He was the Cruise director on board - he may also have been the Captain too, I can't remember. What I do recall is that Boltman was anywhere and everywhere all day and all night long. I just remember wondering whether the guy ever slept. He made guests laugh, sang with the various bands, carried drinks, made a kids scraped knee feel better and helped old ladies up the stairs. Probably to take their daughters dancing that night.

Whatever the case may have been, Robin and I have worked together and played together. He is one of a kind, as I'm sure this prose will be. That's just the way this guy has been put together. Pity we can't say that same for the vessels that he sailed on.

Enjoy every word.

Arnold Geerdts (Supersport Television Presenter)

Publisher's Note: *This foreword was edited to shorten the book.*

A LITTLE FOREWARD
by Ian Sinclair

In a long and distant past when Robin and I were very much younger than we are now, we met aboard an Island-Hopping cruise out of Durban on an ageing MV Aegean Dolphin. This was my introduction to lecturing on the ocean waves and Robin eased me into it by way of a routine of comedy that had my sides splitting most of the cruise.

We would sail together on many magical cruises and my family, especially Kiera, were all smitten by Robin with his genuine warm friendship and his irrepressible sense of fun. His stage routine was never to be missed. It has been poorly copied by so many. So many people remarked that here was the South African version of the legendary Tommy Cooper. I had the pleasure of seeing Cooper live and Robin was better.

I was permanently based on the Oceanos as lecturer and just left the ship for a short break before her demise. I heard the news in a remote jungle in central Africa and was glued to the Beeb for days. I was heartened to hear Robin and the other shipmates being interviewed after the rescue. To hear that Robin had played a significant role in the rescue came as no surprise and his interview on the listing bridge whilst releasing the captain's canaries is one of the greatest reality bites ever.

Over a few drinks in Robin's hotel in the Natal Midlands and viewing his extensive portfolio, I seriously suggested he document his hectic and glamorous life; and just a few months ago I received a disc with a fully written manuscript and images. It kept me enthralled for hours and evoked so many great and emotive memories.

It has been an honour and a privilege to have had Robin as a friend over the years and his publication is a landmark in maritime history and side splitting laughs.

Ian Sinclair - The Birdman

PROLOGUE
Do these stairs Go Up Or Down?

What sort of title is this? - You might be asking yourself. Well, it's one of the many silly questions that a lost passenger might ask you on board a ship. There are loads more, but you can read about them later.

This book is also dedicated to the memories of Tom Hine and Terry Lester, both great friends and mentors who were instrumental in convincing me to become a professional entertainer. Even after their deaths, they continue to influence my everyday life.

I have lived all over the country and spent many years living on the Ocean. Loads of people have come in and out of my life and this is not a time for name-dropping, or anything like that. I'm just going to introduce them to you as you meet them, and the various places along the way. It might even be one of you, or in one of your favourite places, along this journey with me. PLEASE... you don't have to remember them, I'm not going to ask you questions later.

Ian (the famous Bird Book's author) Sinclair and his wife Jackie, were visiting me in my Hotel in the Natal Midlands. After supper one evening during the drinks, laughter and reminiscing about our various voyages together, he said I should write this all down. Ian has written loads of books, but I don't even know where to begin.

I'll just talk and lets see what happens...

Ship
Happens!

SCHOOL DAYS

I was 15 years old when the "Magic Bug" bit me. After watching a magician one day, I decided I should also learn some Magic. I visited every library that the greater Cape Town had to offer, and met some of our local magicians, like Harry Ackerman, Desmond Teale, Vaughan Leader and Brian Marshall. To top it all, one night at a school concert, my prefect, Wayne Abrahamse did a Magic Show at school during a variety evening. During the years ahead, they all helped me and eventually I joined the Cape Order of Magicians. Wayne came to visit the school a few years later in his Naval uniform and I informed him that I was also learning Magic. His reply was, "You are mad enough!"

School days were quite fun. I enrolled in every play or concert that came along, did comedy routines in some of the classes, esp. Mr.Smeda's (Dave) and Mr. Light's (Vernon) classes. After one episode of much laughter, Mr. Light made me stand outside on the window ledge (on the first floor!) He explained as he removed my shoes and socks in case I slipped, that the headmaster might walk around the corridors… if I was outside the classroom door, I might end up in detention. Unfortunately the shop owner across the street telephoned the school and told Mr. Duminy, the head master and told him that there was a boy standing out on the ledge! Mr.Light and I both went to detention that afternoon. We became great mates after that.

Mr. Ross, our vice-principal also loved to tell stories. He was a Spit-Fire pilot in WW II. I was often the instigator to make him digress from the lesson and go into one of his stories. Morgan put his hand up during one of them and said, “Sir, I don’t think its fair that Robin can eat and we’re not allowed to.” Mr. Ross looked at me and I had the largest hamster type cheeks stuffed with Wilson’s toffee’s. He saw the funny side, but I still had to stand in the corner in the waste paper basket/bin. I stood in that bin so many times my school shoes looked like Charlie Chaplin’s. As most older kids we all tried the big walk. The 10th of October, "Kruger’s Day," the then oldest organised race in the country took place in Cape Town. You would get your number and register at Lemkus Sports and walk from Simonstown to Cape Town. The real walkers would do Cape Town to Simonstown and back. A total of 50 miles… 80,4 km’s!

I did two of the *little* ones and when I was sixteen, and trying to get fit for the approaching army days, I decided to do the 50 miles! I gave up uncle Barry’s lift to school and aunty Jeans (Morgans mom) lift home. The big day came and I was up at sparrows. I walked to the Grand Parade (that’s already about 5 km’s) and the Mayor fired the starter pistol at 5 am… I got to Simonstown where everybody else was still waiting to leave Jubilee Square at 10 am, and the announcers had a record of everybody in the race. There were loads of cheers from the Public and friends as my name was called out as I was passing through as the youngest entrant this year. It took forever and I only stopped at the public toilets outside the Newlands Swimming Baths for a leak. My legs were shaking and I was exhausted. I said to the ambulance drivers that had been following me from Wynberg, to put me in the back and give me a lift.

He said they where told to follow me and keep me going. I decided not to quit and 13 hours after I had set off from the City Hall, I finished the Big Walk! At the school morning assembly Mr. Duminy informed the school that everybody knows how mad I can be from the various antics I got up to and the school concerts, etc. But to walk 50 miles you must be madder than I thought, but the school is proud to have the recipient of the “Youngest to Complete the Big Walk” floating trophy. It was nearly as big as me!

I also learned a lot during the woodwork classes which stood me in good stead for my later years. Metal work was a whole different ball game... Our teacher, Mr. Lerm and the accountancy teacher “Mudguts” le Roux were good pals, with one thing in common. The “coffee” in Mr. Lerm’s flask! I also discovered during metal-work the handle for the three different files, the bastard (rough one) medium and smooth became a bit worn after changing them all the time.

We’d nip next door the wood work class room and get wood shavings and jam it into the handle, and was raring to go again! I forgot to do that. Mr. Lerm was on about the second flask of the day and he was standing in front of me giving Morgan a hard time, with his back to me, I started miming a sword fight with my file in hand. Parry, parry, thrust... the file flew out of its handle and torpedoed into Lerm’s back. He turned and saw my empty file handle still protruding in my frozen, shocked pose.”Bastaarrrrd” he yelled and I replied “Smooth!” The class were frozen into position while I raced around trying to escape the Madman. I was eventually cornered, and dealt with... my aching guava!

Then the dreaded day came... my call-up papers arrived. This country used to have a compulsory Military Service, and it is sorely missed in this day and age. What with trying to concentrate on Matric exams, girls and magic, I passed - but only just! Observatory Boys High has since closed and is now a creche for children for the staff working at Groote Schuur Hospital. It was the only school I think where, from the tuck-shop, you could buy a pie, milk and sweets and have a tattoo done while you waited! (just kidding - they didn't sell milk).

All boys and I think I was the only virgin. Just my luck, the year after I left, my school amalgamated with Observatory Girls High. The pupils gave it the nickname Mowbray Maternity Home! Must be Gavin Wards influence with his "Maths Books" - his terminology for "Adult Girlie Magazines!"

Halfway through Matric, the government in their wisdom, decided to change the call-up for Military Service from one year to TWO years! Aaaaaarrgh! Now it said January 1978 to January 1980... that's a lifetime... or so I thought at the time. I was leaving Cape Town to go to Potchefstroom even before the Tit Fairy had arrived, and by the time I got back, they would all be second hand.

The Matric Dance came along and I didn't have a date. David had failed the last year but the standard nines would be there as waiters (I don't know how he failed because we all copied from Morgan.) I borrowed David's girlfriend, Loretta Olney, so at least he had his girl there! Halfway through the night our geography teacher, Mr. Torre came to our table and said, "I know you guy's... where's the drink?" So we took him out behind the woodwork room and finished the rest of our bottle of Old Brown Sherry. The party was a lot better after that!

I'm pleased to say that I'm still friends and in contact with so many of my Junior and High school friends... thanks to the power of Facebook and Instagram, etc.

My last Christmas and Mom's wonderful lunch at home before we all left for the army. My Christmas cracker contained a hat and a little army tank! Playing guitar, sing-a-longs with Uncle Barry and Uncle *Tronnie* - and while the other kids played outside, I was learning magic tricks, playing John Denver songs on my guitar… those carefree days were all coming to an end.

THE ARMY

Armed with a few tricks, we all marched bravely off to the station to platform twenty-four from the Castle... hundreds of boys and thousands of pimples. I shared a compartment with an old school mate called Jack Atmore... The "Rooi Hell" as he was nick-named by the instructors (Bombardiers) later on because of his red hair.

Tears from Moms and girlfriends and all I could think about was that the Tit Fairy had been in the night... I couldn't believe we were leaving all these girls for the guys that were only being called up in the July intake. Now we had to make new friends and discover a whole new life. How will you adjust to new surroundings, people, rules and realising you had left your childhood on the train? I don't remember much about the train, but the musicians along the way begging for coins all knew the same song - no matter where we stopped, or what time of the day or night. *"Daar kom 'n hoender aan, hy het sy bloomers aan; Wat sal die ander hoender's se?"* (translation from Afrikaans basically is... "Here comes a chicken, *He's* wearing her bloomers, What will the other chickens say?") It never quite made it into the top twenty! I do know that most of the hard green sausage shaped cushions didn't last the journey... most of them were thrown out before De Aar... along with our schooldays and childhood!

We saw our first mine dump on our last day on the train. Our new home was getting closer and the Army Bedford's were idling and growling, just like the instructors, waiting at the Potchefstroom station. We were all very quiet and the instructors were making all the noise. We were piled into the trucks and *Roofie* rode to 4 Field

Regiment, hitting every bump and curb the driver could find. I made a few good friends on the train but I never saw them again after the haircuts... everybody had dramatically changed. I could only recognise Jack (Redhead)... I slept in a tent on the first night for what seemed like just a few minutes... at about 4-00 in the morning this awful voice boomed out shouting "*Tree aan*!" (Afrikaans for get together, your world is about to change as you know it.) I remember Jack saying, "*F*** - it's still yesterday*!"

We all collected our kit and we marched all over the show. We ran to imaginary trees on the horizon and did about 4000 push-ups in the first few hours of the day and the sun had only just crept over the horizon.

The medical was quite embarrassing... hundreds of willies of different shapes and sizes... I managed to fill up my little beaker from my little willy and I was declared fit for Military Training - G1K1! My thoughts of maybe going home earlier were dashed. I would have to endure this awful nightmare for the full two years.

The days just went from one into another... Gunner R. Boltman even had to shave and I had nothing on my chin at all. Some of the guys got their five o clock shadow at about lunchtime. We were introduced to the 25 lb. guns and I asked the Bombardier as to who fires these things? " You will soon," he said. Us! We can't fire these things, we're still boys... I was only seventeen years old!

Pat Kerr kept us entertained on Springbok Radio with a show called Forces Favourites. We would listen to the requests for the boys in the forces, on the border or still doing basics in camps - love sick girls with the regular message... "to the army you're just a number, but to me you're everything." Watching our washing dry (I forgot to take a chain and a padlock to stop my washing being nicked) learning to make a bed that looked like it had never, or will ever be slept in again, washing, ironing, sewing, etc... we were learning to become real men!

One night we went out into the middle of nowhere to learn how to rough it and how to creep around in the dark covered in "Black is Beautiful" (a black camouflage grease that takes about a month to wear off! It probably would be banned in this country now!) We arrived back to base and everybody looked quite pretty in our new found eyeliner that we couldn't wash off.

We were quite excited because we were told we were being paid today... I didn't have to wait too long because it was done in alphabetical order... Nobby Alborough was always first... I signed for my R30 which was then immediately decreased by deductions for various causes (hair-cuts, tomato sauce, NG Kerk, etc.) and I ended up with the grand total of R27-50 for my first wage packet ever... and then to my shock and horror I saw the date next to my signature... 31st January!

I had missed the first half of my 18th birthday. I was now legally allowed to keep the R1 Rifle I had in my *Kas* (Cupboard). Bungalow 391 was run by Bdr. Fourie, probably the only real gentleman and soldier of the lot. The other firm favourites were some of the officers and permanent force officers. I had quite a good time because of the magic I could do, and the sing-a-longs with the guitar.

One day the siren went off. Veld (field) Fire! Everybody went to the parade ground in the dark to get onto the waiting Bedford's to be taken out to fight the fire. Lindsay Hunting and I climbed on the roof of the mess. We weren't going anywhere… Margaret Gardiner was going to represent South Africa in the Miss Universe competition and we didn't want to miss our Capetonian. When the dust settled we sneaked back to our bungalow and switched on the only TV set which we had all clubbed together for from our measly R27-50 a month and sat down in the dark to watch.

Suddenly there was movement in the darkened bungalow… we watched wide eyed as a *kas* door slid open, a *Trommel* opened, Lloyd Brown climbed out of his water-bottle and people started appearing. From various hiding places, all the Capetonians came out of the woodwork to watch. Even some of the bombardiers pitched up. Bdr.'s Gouws and Joubert were also from the Cape. We weren't popular when the lads came home black from fighting the fires in the wee hours of the morning and we all had *"tents in the bed"*… Margaret had won!

We were pulling out weeds one Saturday morning in the pouring rain. Bdr. Erwee was in charge sitting in the comfort behind the wheel of the open backed Bedford. We had to fling the weeds with their soggy wet roots of mud onto the back of the Bedford. He was shouting the orders out of the open window… they all seemed to be qualified in various trivial things like weeding, digging, etc. Well, you could see it coming, can't you? Nobody knew who flung that Bull's Eye weed, but it landed right on the side of his face. He swore every word, even words I hadn't heard before… in both languages! We were going to pay dearly for this. Our entire bungalow had to be present after supper in our bungalow. The hour drew nearer when suddenly the bungalow door flew open. We all jumped to attention and a much cleaner Bdr. stood in the doorway.

Neighbouring bungalow friends came to peer through the windows… we hadn't experienced bungalow PT before. First we had to pick up our trommel, hold it until our arms grew by two inches, then put it down. Then pick it up, then down. Then hold your rifle parallel to the ground out in front of you with both hands and begin turning it like you were winching something… I felt like I was winching up an oil tanker. Then the push ups and the sit ups. Then the fun started… climb in your *kas*, then get out… a few of our guys could hardly fit into their beds, let alone their little "wardrobe." It was okay for some of us who were still growing… Lloyd and I could do a three point turn!

The windows by now were somewhat steamed up, and our neighbour's were battling to watch the show. Then we were ordered to face to the right and crawl as fast as we could in a circle around the bungalow. It was like an obstacle race… over a bed then under the next one, over and under. By now of course we were laughing like naughty children, and the more we laughed, the more annoyed Erwee became. We would also pause to examine some of the parcels that were stored underneath some of the beds. The rich guys were always getting parcels from home with loads of goodies. You could hear them at night trying to open packets with the least amount of noise - a bit like unwrapping sweets in your school blazer pockets. Some of my friends had a degree in this department. Anyway, it was a free for all, and you left tasters for your mates that were hot on your heels. Screams of discovery and screams of horror as various guys found their favourite nosh, or discovered their hoard had been Noshed! Eventually Bdr. Erwee called it a night and we all went to the showers, still laughing. It was probably the best keep fit class I've ever attended. My arms are still longer than most of my sleeves.

Lindsay Hunting and I decided to hike to Cape Town on a weekend pass, because Graham Inder was going to fetch his car. We had a guaranteed lift back so off we went. I've never been so cold in my life. It took us for ever and we only had a few hours at home... I wanted to surprise my friends... I got the surprise... most of them had gone camping for the weekend!

On the way back we stopped to pick up Lindsay in Stellenbosch and he piled into the VW Beetle carrying a strange looking wooden suitcase... Bagpipes. What an entrance we made when we arrived back in the wee hours on the Monday morning. Sleepy heads were brought out of dream land to the sight and sounds of Lindsay marching down the bungalow with a screeching tartan cat under his arm. After that, wherever 42 Battery marched - Lindsay led the way.

After Basic Training we got the news. Our 42 Battery was going to become 43 Battery - we were being transferred to Walvis Bay in the then South West Africa - now Namibia. I had established myself as a comic and magician and we did a few performances for the Regiment before we left. Cat, aka Christopher Austin Tennant was another character who used to do a wonderful version of the Rocky Horror with me on the guitar… what a sight, in army boots and browns strutting on stage doing Frankenfurter. We had a good send off, and a few days later we headed for the station en route for Walvis Bay. The green sausage shaped cushions of the South African Railways didn't make it again. The minstrels were en-route again… *"Daar kom 'n Hoender aan, hy het sy bloomers aan."* After a few days on the train we arrived on the West Coast of Africa to be greeted by sand… loads of sand!

We were piled into Bedford's and taken to Rooikop… our new home. En route we were introduced to the notorious Dune 7, apparently the 7th largest shifting sand dune in the world. I remember saying to my wide eyed companions that we had better f*** off before the cement arrives. We were going to go up and down on that pile of sand many, many times. I think I learned more swear words on that pile of sand than in any bunkers I'd be landing in on golf courses later on in life.

In Walvis we met a staff sergeant called Hook. *"Klim Op en Klim AF"* was his major greeting whenever we went anywhere by Bedford. Three stripes on his sleeve with a volume control above it. It only turned to the right! He was watching us in the food line one night, while we were out in the desert doing a night shoot. The wind was blowing and there was sand everywhere. As we got closer to the hot boxes, in the lights of the Bedford's, I said to the Sgt. Major - *"Good evening, can we have a table for four in the smoking area please?"* The press-ups were well worth the laugh!

The camp in town was always covered in mist (or so it seemed) but it was nice to be able to walk into town on a pass. When the *ou manne* left, we moved to *Rooikop.* It is out in the desert away from the rust at the coast and named after the rocky outcrop of rock that's also handy for the bombardiers to chase us up and down. We visited the top of that rock many times.

I used to often wander off in the evenings on my own and find a quiet spot to play my guitar... Gary Frayne would sometimes sneak up and listen to some of his favourite songs - he was also a John Denver fan.

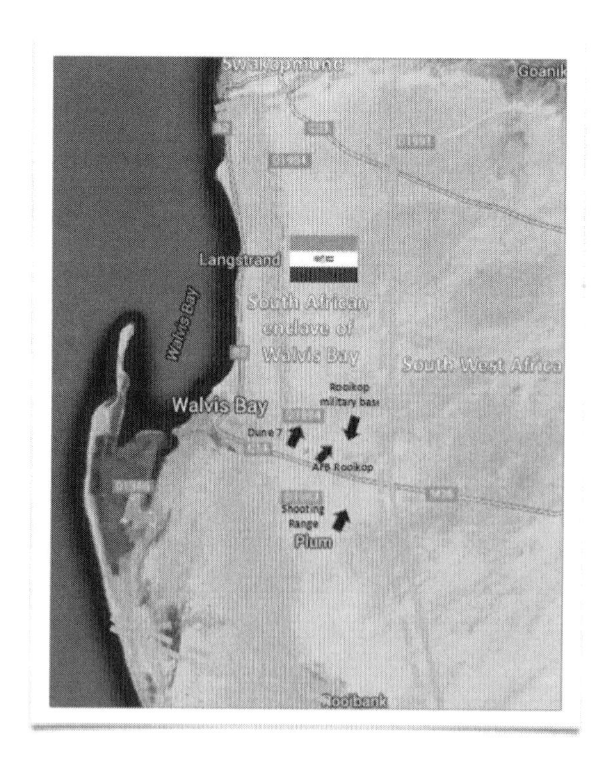

THE BORDER
Katima Mulilo

We were being trained for the border - half the guys would stay in Walvis while the other half of the battery would go to Katima on the Caprivi Strip. I couldn't come to grips with quite a few things that the *Powers That Be* would do. We looked quite silly out in the Namib Desert with camouflage nets above the guns - it looked like eight square trees in a row. It did make sense however when we arrived in Katima Mulilo which is in the most beautiful jungle. Wild life was very evident in the area, and we could hear the Hippos and often saw elephants in their natural state. We all wore browns and the police in the cities were in camouflaged uniforms?

Major Human, *Papa,* was the boss with two Sgt.Majors, *Viskop Visagie* and *Nine Mil* (He always had his 9 mm pistol on his hip). We all became quite fit from the squash court, the Rooikop rock and Dune Seven. We also used to go into town on Sundays to church where the ladies of the congregation used to make us tea and cakes. A Cape Town minister had been transferred to Walvis Bay. Rev. Mark Stevenson proved to be very popular and the little Methodist church had to go through some alterations to accommodate his growing flock.

His lovely wife, his red VW Combi with surfboards on the roof racks, his long blonde hair, he looked like anything but a minister. (What's a minister supposed to look like?) He gave a concert in town involving some of us and he also played the guitar. We became quite friendly and he came around to our base to say goodbye when we left for the border. Our food was delivered to us from the base in

Walvis and arrived to us in hot-boxes... thirty kilometres later the boxes were never hot. One Sunday the food smelled off and I decided to let the chef know by sending them back with messages on the hot boxes. The next day the shit hit the fan. The bombardiers arrived at sparrows, got us all out of bed and we had to wait for the Major.

He arrived fuming and *Viskop* got us all to attention and asked the guilty party to step forward. I did and was drilled at an alarming rate into *Papa's* office. Of course when I tried to explain using my hands to gesture, I was smacked on the back of my head by the out of breath *Viskop,* for moving while standing to attention. (He was not used to marching at that speed either) I said I can't talk if he's going to hit me all the time, so he was dismissed and I was left alone with the Major. I was lectured on the correct procedures to follow up the ranks, etc. I said that this was the earliest any of the instructors were ever on the base and I bet him a case of beer that we would have a lunch today that was better than we have had before.

We shook hands and he gave me the tank track which he kept in his office and sent me off to the top of Rooikop. I went jogging past the guys who were still in formation and Peter Fortune, Peter *House hammer,* Gary Frayne, Lloyd Brown, Hillary Dollman, Peter Clouse, Noel Slabbert, Carl Spires, Chris and a few others joined me and off we went up to the top singing the Battery song. That spirit didn't go unnoticed.

At lunch time the RSM and other brass arrived from HQ and we had a great meal. The Major stood up and got everyone's attention and thanked me for a lovely meal. The guys all applauded, but he wasn't quite finished - because I seem to know so much in the food department he informed all of us that he had now made me the chef for when we arrived in Katima! (I lasted about a day in the kitchen as a chef - "*What shit is this*"? comes to mind - Lloyd Brown took over)

That evening around our canteen area the Major came down and presented me with the case of beer and we all had a few drinks, including the Major.

Katima Mulilo in the Caprivi Strip is a beautiful place with loads of animals, trees, birds and bigger guns that we had been trained on, to get used to. The 140 mm or 55" as we called it was certainly a great weapon. Eight of them, all under camouflage nets - now we knew why from the desert training - all facing north. Golf Base (Golf for Gunners) had sand bag lined bunkers which we all slept in. A large bulldozed wall surrounded the camp with trip flares, claymore mines and guard bunkers armed with Bren guns and Browning's.

Two huge trees with seats in them so that we could keep watch over the wall during the day in case any terrorist crept up to the camp. Guard duty every night from sunset to sunrise, vehicle guards to and from the town, morning patrols and kitchen duty. We all became remarkably brown from the tropical sunshine when a gun crew had the odd day off. We became even fitter with sandbag PT, projectile PT and flipping water-filled tyres. My mouth always got me into trouble and I can't remember how many dustbins I scrubbed into mint condition. Carrying 100lb. projectiles wasn't fun.

I remember telling one of the instructors that if my mother knew I was running around with bombs they would all be in big shit - more f***ing bins!

We only saw action once. Apparently someone from across the river in Shisheke in Zambia fired on our patrol boat. The camp OC, Capt. Bosman (ex 42 Battery in Potchefstroom) decided to get the gunners into action. Eight guns fire for effect. We rammed the projectiles home after writing Christmas messages (Lots of Love Peter, Chris, etc.) on each one and then fired on the little town for about half an hour. It was Boxing Day.

Chris Tennant (above with sunglasses) and I were “fired” for being trouble makers and we were sent back to Walvis. A great reunion was had with the other half of the battery that had remained behind. There was method in their madness - we helped inform our mates about the border and assisted the instructors what to expect during their lessons. Two weeks later we were back in Katima with the replacement battery. I spent about 13 months on the border. I was often loaned out to other units in the area to perform. It was interesting seeing how some of the other guys were living.

I made some good friends in the army and I’m pleased that some are still mates to this day. (Peter Fortune named his first child after me) We all survived the border but were nearly killed on arrival in

Walvis when the landing gear didn't come down on the *Flossie* (Hercules). They aborted the landing at a hell of an angle - through the hangers and buildings and then landed safely. We got off and everyone had instantly lost their tan except for our underwear! Gavin Ward, a school mate, was in the Air Force there - he was on guard duty at gate when we drove out, *"Did you land or were you shot down?"* he asked me, sitting in the front passenger seat.

Our 40 days party came and I was detailed to go out into the desert to collect firewood. (I know that sounds strange, but there are areas where there is a bit of growth and in this area Mother Nature has a wonderful canvas)

I had been learning to drive secretly (thanks Stuart Hopwood and the Unimog *Astrok* on the border) and one day whilst parking a Bedford, I accidentally reversed into the hanger and damaged the door and the Bedford. The drivers were very helpful and gave me a number of another Bedford that was in perfect condition. I prayed that the rank wouldn't notice the new wood shining through from the door post. Upon my return with a huge pile of firewood, I was sitting proudly behind the wheel and was greeted by an irate *Viskop*. Shit, *they had discovered the door!*

I was drilled again into the Major's office at high speed. The major put his arm around my shoulders and walked me towards the broken beam. I'm sure the broken exposed piece was getting brighter with every step." Why didn't you tell me about this" he asked. "I'm quite fond of living," I replied. I think I was still trembling when we inspected the immaculate Bedford. It was fine but I was told to stay out of the driver's seat of any of his vehicles. Off I went up to the top of *Rooikop* for what was to be my last time.

The 40 day party was great and I am sure we were all done before the meat was. To this day I can't handle the smell or taste of Coco-Rico (Malibu). At the Waldorf restaurant in Walvis Bay there was a jukebox, but if you weren't fast enough with your selection it would automatically play Cliff Richards' Forty Days. We were given the good news that we will be home for Christmas. I collected money from all the guys, bought decorations in town and one Sunday set about decorating the huge pine tree that stood out in the garden. (Only the Army can make a pine tree grow in the desert.)

When the PF's (Permanent Force) arrived, the Captain made me take it all down. I was halfway done with the task when the Major arrived and said what a brilliant idea. No other in-take had ever done that. From halfway up the ladder I said that the Captain didn't think so, I'm busy taking it all down. He bellowed for the captain who came running towards him with that, 'what have I done now' type of look on his face. The major sent the captain to get all the remaining battery funds, drive me into town and buy Christmas lights and help me put it up. That evening we all sat around with the guitar and the major turned the lights on. We sang Christmas Carols and John Denver songs, much to Gary Frayne's delight, (National Roads, take me home - to the Cape where I belong. Cape Peninsula, Table Mountain - Take me home, National Roads) into the wee hours, after the captain had been sent down twice to refill the beer bath with the remaining funds.

I had 'borrowed' a sheet from the stores and with nugget and a brush, I drew Table Mountain on it to hang from the train which would take us all home. The Major came along to say goodbye and proudly signed the sheet and I detected a lump in the tough old bugger's throat. I would see him again a few years later during a camp.

He was a Colonel then and I was asked to entertain the top brass at Army Battle School in the Northern Cape. The RSM came to collect me and drove me to the venue. I was still a gunner and was made to stand at the door while the RSM stamped to attention and saluted. The Colonel had his back to us, but I recognised him and when I shouted Papa, he swung round! The RSM was horrified but the Colonel came over and we hugged each other like long lost friends.

I don't remember much of the train ride home other than the fact that the green sausage shaped cushions didn't make it again. They must still be buried under that dreadful Dune Seven. It was quite emotional at De Aar when we parted company with the Natal and the Transvaal boys. To think that we hardly knew anybody when we were sent to Potchefstroom two years ago.

We had been living together in good times and bad - some laughs and some sad... "*We'll meet again, don't know where, don't know when, but I hope we'll meet again some sunny day.*"

Then we set off to the Fairest Cape and the same minstrels were still serenading the troop trains. "*Daar kom'n hoender aan...*" Now we had a new way of life to get used to. Finding a job, earning money, choosing a career? Well, we'll tackle that after Christmas and New year - we're civilians and we need to see what the Tit Fairy had been up to in our absence.

ALL IN A DAYS WORK

My first job was with the Department of Sea Fisheries. I had always wanted to go to sea and I didn't know then that in a few years it would become quite a part of my life. The closest I got to sea with them was the harbour wall in Kalk Bay! Dressed in a white lab coat, gum boots and jeans, I was responsible for measuring or reporting undersized catches and to take stomach samples of fish from the different areas of False Bay. I didn't know a stomach from a liver! Tourists would even take photos of me! What a job! I didn't last long there. The two Harbour Masters, *Vossie and Pottie Potgieter,* said I didn't belong here, as they handed me my measly salary. I left because the salary was like PMS - came once a month and lasted about a week! I was to learn later in life that this phenomenon can actually last a lot longer!

Life long friend Mark Rauch got me to apply for a job at Nedbank where he worked. I enjoyed the interaction with the staff and the clients. I stayed with them for five years. To this day I still remember account numbers and clients, but the staff were great. Jack (Rooi hell who you've already met) came to work there. Marks sister was also there, Sharon - what a beauty. The Fairy was very good to her! But enough of that - she's like family to me.

One day a stranger came in - February is as hot as hell - and he came in with a coat and rolled up balaclava. Very suspicious indeed! He left when the manager came out of his office. I asked Jack who was on the counter (teller) at the time if he had asked for anything, but Jack said no. He came in again and I wrote some details down about him when suddenly he tried to grab a bag of wages that

Swannie was making up. Swannie swung the bag at his head and the would-be robber pulled out a gun. *"Down"* I shouted and all the tellers disappeared. I ran over to the accountant's desk and took the revolver out of a very wide-eyed Derek Munting's drawer. I raced towards a now fleeing robber who dived into the getaway car and I stood in the road in Salt River firing into the back of the car, just like Dirty Harry. Brian Brodrick came out of the back door armed with his own firearm as back-up. They were caught later in another foiled attempt in Hout Bay. I went to the prison to an ID parade and unfortunately I identified a warden! They had had their haircut. Anyway they sat for seven years and I got a letter from Nedbank shitting me out for taking action because of insurance issues, etc. and in the second paragraph thanked me for my bravery!

Before Christmas one year I caught a bag snatcher. I was on lunch upstairs and I heard screaming from outside. There was a big black lady trying to wobble after a young guy, which in *those* days, looked very out of place racing away with a hand bag. Anyway I ran down through the banking hall and chased him all the way to Salt River circle and retrieved the hand bag. Luckily he didn't put up much of a fight, because I was knackered just from running. The fit Army days were gone. Back at the bank I had to go home to change because I had his blood all over my shirt. Upon my return there was the big black lady who came to thank me and all the clients applauded as she gave me a big slobbering kiss.

Speaking of changing clothes, I came to work one sunny day dressed in a sky blue shirt, white tie and white trousers and a few clients remarked how refreshing I looked. (Nedbank was a very staunch, strict bank in those days) The manager, Tom Naude, later came out and asked if I was selling ice-cream or playing cricket?

He sent me home to change and I returned wearing my tuxedo, bow-tie and all. I did magic at the counter, the staff were laughing but Mr. Naude didn't talk to me for days.

Magic was taking over and I was doing a lot of moonlighting. I met many entertainers in the various night spots in Cape Town. As I was a member of The Cape Order of Magicians I learned a lot from Vaughan Leader, Brian Marshall, Wayne Abrahamse, Rowland Hobbs and the very young Wolfgang Riebe (now a big name in Magic and a good mate).

Entertainers like Rupert Mellor, Butch Cook, Joe Parker, Derek Gordon, *Late Final* with Richard Black and Jerry Barnard, Richard Hyam, Pendulum with Rusty, Derek and Linda Gordon, Abbott & Crabb, Josh Sithole, Dave & Harry Monks (Leprechauns), The Danny Fisher Road Show, *Altitude* and Jeff Weiner & Mainstream were always inviting me on stage.

The Cape Order of Magicians did a show at the Labia Theatre for a week. I used to do the rabbit out of the hat trick, but Rowland Hobbs was on before me, and each time Rowland's music, *Shakatak* came on, my rabbit knew he was about to pulled out of the hat... he used to go around the bend. During one performance the rabbit suddenly appeared at my feet beneath a table. A little girl yelled out, "Look mom, a rabbit!" I was horrified - my last trick was on the stage and I was only halfway through my act.

Animals are always unpredictable. I opened the curtains during a performance with Wayne Abrahamse one evening to let the audience have a good laugh... Wayne was crawling along a beam trying to attract his doves back into their basket! On his bachelor night we eventually ended up at the Sportsman's Bar in Newlands and Butch Cook invited Wayne up to do magic on stage with a ball and chain attached to his wrist!

It was about this time I started to progress from children's birthday parties to adult night club shows. Steve Albert booked me for his spot called the Coimbra in Observatory, just below the famous Groote Schuur Hospital, where Prof. Chris Barnard performed the first Heart transplant. Later I was to work at Steve's Restaurant in Claremont, The 'Rareside Grill' for over a year doing Close-up Magic. He now owns Nelson's Eye in the Gardens. I did Club 604 Ladies Nights for the Martin and Dave Rattle brothers and also Charlie Parkers for ages.

One day Derek Gordon (of Naughty Lemon fame, and the rest) and I were booked for a gig at the President Hotel, Raffles Bar for an exclusive men only function. Bearing in mind in those days nipple caps and G-strings were the order of the day. We did our bit and retired to the bar and watched the girls strip on the dance floor. One came on dressed as a gorilla and slowly started to de-fur herself.

The next day I got a call from the police queering the shows contents but I said I'm a magician and I don't know what happened after I had finished my bit. They phoned Derek and he said a similar thing and then elaborated about the gorilla... *"This gorilla started molting,"* he said, *"except for a little triangular patch!"* Everybody knows Derek Gordon - who would argue with him? (Except for his cell phone - he is always having hassles trying to get the thing to work - as he explains in this new era of modern technology... "I'm as useless as a horse on a surf-board!") I'm a bit like that also -

children talk about Apple iPods, MP3's, etc. In my day if you had a three inch floppy, you hoped that nobody else found out about it!

It was round about this time I met Johnnie (Biltong & Pot Roast) Noble, Tommy Morton, Barry Hilton and Peter Hughes. I was a bank teller with Nedbank and Barry was an electrician for the Arthur's Seat Hotel in Sea Point. Johnnie had opened up an entertainment agency and I remember one evening at The Lady Di bar with Kenny Oliver, we all did a spot and were paid the grand sum of R80-00 each. Barry has since become one of the most popular comedians in this country. He phoned one evening a few years ago and said that Jane had just given birth to another baby boy and we want to name him after you. I said you can't have a little boy walking around called Boltman. That little Robin is now bigger than me.

At the Hout Bay Hotel on Saturdays, Jeff Weiner & Mainstream entertained the masses. All my mates and Tracy Johnson couldn't wait for Saturdays. (Or the gorgeous Gabby Bianchelli) They all used to come in with me to avoid being charged the entrance fee. I tried to do something different every weekend because the crowds were nearly all regulars. I learned a lot about working a crowd like that - I learned even more in the car park in the back of David's red Granada! I'm not to sure how many love-affairs started and ended at The Hout Bay - sometimes all on the same day!

That's where I picked up the nickname - the *Hout Bay Magician*. I was later to make headlines in the Sunday Times travelling the world with the notorious Hout Bay Passport. You had to put your own ID picture into the realistic looking passport above a caption that read... *The bearer of this passport is entitled to any country with a sense of humour.*

SA's travelling trickster!

COMEDIAN USED A HOUT BAY 'PASSPORT' TO TOUR THE GLOBE

By IVOR CREWS

A SOUTH African prankster conned immigration officials in several foreign countries with a Republic of Hout Bay "passport".

"It was done as a practical joke," revealed magician and comedian Robin Boltman, 30, of University Estate, Cape Town, this week.

He used his Hout Bay "passport" to get into Fiji, Australia, the Ivory Coast and New Caledonia during a world cruise on the liner Achille Lauro.

"I topped it by clearing the strict immigration procedures at London's Heath-

OUTLANDISH ... Robin Boltman and 'passport' Picture: TERRY SHEAN

There were so many venues for live entertainment in those years. From the Hout Bay, Sportmans, The Pig & Whistle, Club 604, Charlie Parkers, Raffles, Blazers, Rompies, the Oyster Bar, and downstairs The Whistle-Stop/Pool Bar, The Liz, The Clifton, the Ambassador, Inn on the Square, Square One and Peabody's, Naughty's, Chelsea Arms, Charms, The Holiday Inn's Samantha's and Lilly's Bar, Prince of Whales, the Canterbury Inn and the London Town Pub and upstairs at the Century, the Crazy Horse, Seagulls and the Villa Revue. At Basin Street Blues I had the pleasure of performing with Rupert Mellor on the opening night.

The place was so full that Rupert stayed seated at his piano and I performed standing next to him. I finished with the torn and restored newspaper trick and afterwards produced a live dove from the restored newspaper. Rupert said, "F*** it!" and it came out over his microphone. That also helped getting the applause. The Stags head was another one I popped into one evening after a show at the Oyster Bar to get away from the maddening crowd, and heard a great musician in action. He then played a song claiming it was one of his two claims to fame? He started playing "Where do you go to my lovely" - I had the pleasure of meeting Peter Sarstead! (His other one was "Frozen orange juice").

The Godfrey Ulster Trio in the Cape Sun's Noon Bar and the Chapman s Peak Hotel. I also performed in places like the Athlone Hotel and worked with some of the finest entertainers that Cape Town had to offer. Taliep Pietersen, Tony Schilder and his band (mostly family all of who were multi-talented), Leslie Kleinsmith, Zayne Adams, Terry Fortune, JJ King, to name but a few. Hope I haven't left out any of your favourite pubs, but I do hope I've re-kindled your memory of some of the Magic institutions!

During my Nedbank years I met Terry Lester, Robert Kirby and Maureen England at the Holiday Inn. They used to do a show entitled *Dr. Kirby & Dr. Lester's Beaver Protection Society.* Robert Kirby was a satirist and also a magician and he taught me a trick I still do to this day. After the show, Terry would stay on and do his own cabaret with Tony Schilder on the piano. It was during one of these shows when I introduced myself and asked if I could do a little bit on stage? He said, "*Drop the Mr. Lester - call me Terry!*"

Years later he was to be my MC at my wedding in Leeds! I learned a lot about audiences and magic at the old Holiday Inn. In the beginning it was Pepe dos Santos and Ivano Fattori and *Winds of Change.* They used to invite me up to do ten minutes and the confidence grew. Touring bands from the UK would perform there on contract and move around the country with the Holiday Inn. Danny

Fisher's Road Show, Des Lee, Altitude and Ricky & the Spitfires. Most of them are still living here, after taking up permanent residence in this country.

I also performed on Saturday nights at *La Vita* - a restaurant owned by the enigmatic Professor Chris Barnard. Quite a few Nedbank clients used to dine there, and it was always quite amusing when they used to study me on stage with that - *where do I know this guy from,* look on their face. Out of place from the bank they had no idea... to make matters worse I did close-up magic at the tables and read their minds... like quote their account numbers... then they clicked. It was on one of these occasions that Chris Barnard said to me, *"I wish my hands could still do the magic that yours are doing today".* Two weeks later the Sunday Times came out with the headlines that Prof. Barnard would no longer operate.

I was taking a few days off my leave with Nedbank to do shows and I was asked to pose for a brochure by Rightfords, an advertising company, as a magician for the Safmarine Cruise brochure. Glynnis Delaney (Delaney Marketing) hired me for a weeks gig around the country celebrating Martel Brandy's 21st year in the country. I met the Sedgewick brothers, Christopher (father to my army mate James) and Antony, Jacques Martell and Ronnie Melk and "Duimpie" Bayly, and travelled the country in a Lear Jet, entertaining anybody and everybody that dealt with Martell and SFW.

I took unpaid leave for this one. On the last night it was Jacques Martel birthday. It was held at the Wooden Bridge in Milnerton. Two big shots from Safmarine asked if they could attend the function. They had to sit up at the bar and Jacques made a big fuss of them, pouring Martell Cognac down their throats. After the show they had disappeared but the barman said that my guests had said thank you and they would call me... I thought I had clinched a show for the Safmarine Christmas function. After the festivities Jacques insisted on me taking him home in my Ford "snot green" Cortina, cracked windscreen and all. I first took him to Forries and then back to the Mount Nelson where he was staying. He insisted that everybody at the bar drink Martell Cognac and I eventually bid him farewell when I was speaking French and he was sounding very Afrikaans!

Sporting a hangover from France, I returned to Nedbank the following morning to receive a phone call from Safmarine, asking me to come and see them in my lunch hour. When I got there they said that they had enjoyed the show, followed by the words that would change my life forever... "*How soon can you join the Astor*?" I replied, *"I've got my sandwiches in my pocket. Where's your ship?"* When they asked about doing a trick right now, I noticed boxes in the corner of the office, with Rightfords printed all over them. They were the marketing company that printed the Safmarine brochures. I tore open one of the boxes and asked June Barclay and Stuart Venn to page through until I told them to stop. There I was - top hat and tails with a dove in the Astor brochure and they had only just offered me the job... Magic!

I don't think I ate my curled up, smiling Marmite sandwiches that day. When I got back to Nedbank Salt River I asked Derek Munting for the resignation forms and he said don't do anything rash. I told him that I'd been offered a job as magician on board the Astor! He flew me into the manager's office and said Robins going on the ship. We had loads of drinks that afternoon in our little staff pub.

I still had to work for another month due to work policies. I spent these working days teaching staff members my jobs and my last two weeks I spent painting the vault and re-packing the filing system!

One evening I was out at Forries, and on my way home I was coming around Hospital Bend and I saw that the Astor was in port. I carried on driving right into town and as close to the docks as I could get, because we still had Customs Officials in those days. I stood and stared at this beautiful, sleek Cruise Liner all lit up, and couldn't believe that this was going to be my new home - and a whole new way of life. Needless to say I was too excited to go home for supper, and I went to the Capetonian Hotel where Derek Gordon was playing. He introduced me to Tony Starke from the ship. Tony was the drummer in the Peter Adams Sound. He was also later to become my cabin mate and a good friend too.

The day I joined the ship, 4th November 1984, we first had a gathering at *Forries,* followed by *The Pig* and then down to the ship. My mom and Froggy were on board before me to wish me Bon Voyage; along with a few of Nedbank's staff of 22. I don't think any of the tellers balanced that day.

THE MS ASTOR

I shared a cabin with a singer called Erez Shaked. He was a singer with the late night band, Erezand with John Ford played in the Lido Lounge. Man could he charm the girls! The Cruise Director was a charming girl called Ve Maskell. I met all the entertainment staff during a staff meeting and besides being in the cabaret line-up, Ve wanted me to perform close-up magic in Harry's New York Bar - a popular bar on board and the resident pianist was a gentleman called Tom Hine. We were soon going to become the best of mates - life long friends as it would turn out. I would do the old cigarette in the jacket trick and he would play *Smoke gets in your eyes.*

Tommy Morton, who you've met via Johnny Noble and Barry Hilton, was also on board. He is an absolute nut-case and a real tonic for the crew and passengers alike. On board we had entertainers coming and going, just to add variety to the ship. The core would stay together and head-liners would come on... Zayne Adams, Johnny Tudor, Tony Monopoly, Neville Nash, Des and Dawn Lindberg and Rob Duncan.

One voyage after the staff introductions, Tom Hine and I went into Harry's for a drink. Two passengers decked in binoculars, cameras and whatever else they could hang around their necks, approached us and enquired if we had ever seen any wild life at sea. I said of course - Petrels, Terns, Gannets, Dolphins, Whales, etc. - but not to worry - if anything interesting pops up the bridge will announce it over the tannoy (PA) system. Tom, looking like the ships doctor in his white safari suit... said, "*We saw a mermaid once - just near the Seychelles".* Their eyes widened in astonishment. "*Remember, Robin*?" - I nodded and then he added," *She was a beautiful blonde, sun-tanned... and her figure was to die for... 36 - 24 and R2-50 a kilo!"* I nearly fell off the bar stool. Tom was full of these sorts of quips.

At staff intro's, Terry Lester (he was also on board) never knew what to expect from Tom. As the elderly English gentleman that he was, and being the ships cocktail pianist, he would come out with a Captain Hook type hook sticking out of his sleeve that he had borrowed from the ships butcher. On one occasion I bandaged him up and with pencil rubbing on paper, gave him such a black eye. The staff were horrified, they thought Tom had been in a fight ashore! Tom was a wonderful pianist and raconteur. He always started his set with *There will never be another you (Ewe) - "For all the lonely shepherds!"* *As time goes by* was another favourite (and all these years later, it was also the name of our house. Coincidence?) He would embellish as the song went, "*a case of do or die."* Tom would stop playing and ask the audience if they had ever been into a bottle store on New Years' Eve and asked for *a case of do or die?*

The world will always welcome lovers... "No - my sister came home one night and announced to my parents that she was pregnant!" Then sing - "*The world doesn't always welcome lovers, as time goes by*".

One of the favourite morning activities was a gambling game that was very popular during the Royal Mail Ships' era. It's based on the mileage steamed from 12 noon to 12 noon. Tom and I were the hosts, me for being able to count money and remembering passengers names. Tom got away with it by calling everyone *My little pet,* etc.

On one occasion we had a fellow called Eric Yarrow. On the tote auction board when he had purchased a mileage, I wrote *Sir* Eric on the board... Tom gave a little glance and quizzed me afterwards. I said he is Sir Eric and Tom apologised to him and Sir Eric said, "*I call you Tom and you call me Eric... let's keep it that way.*"

Another fellow we met, I think his first name was Roy, but his surname was definitely Shanks. Tom asked if that was like the Shanks on all the urinals and toilets around the world? Yes, he was on his way to South Africa to see some of his factories and people, and Tom responded that our business was just like his... *Getting bums on seats!* He remained a pal the whole voyage.

I don't know what it is about cruising, other than it can get you hooked, but sex seems to just come naturally. There's nothing like a bit of *off shore drilling!* Is it the sound of the waves, the ribbon of moonlight over the ocean, or the smell of the life-boats? (Or the smell of the smoke coming up from the crew deck!) Because of sharing a cabin these moonlight activities were somewhat restricted.

Tag and Release! My first passenger was Marietta - up in the conference room. A cabin stewardess called Debbie, while Erez was performing in the Lido Bar. Then it was Dawn, a hairdresser from England up on the helipad. She left me with a lovely gift when she disembarked in Lisbon, but the ships doctor soon sorted that out! (*Mucky Bastard!)*

The new hairdressers from England joined the ship and all the guys suddenly decided that we needed haircuts. I fell in love with the manageress, Liz Pask, and we are still friends to this day. She nearly got me fired from the ship. One evening during a party with the crew below decks, we decided to go and get some fresh air, smell the life-boats etc. We ended up in the pool followed by a mad rush into the mens shower. There we were at three in the morning with our clothes leaving a path that the blind could follow... unfortunately the ships patrolling fireman followed it first. I was summoned to the staff captain, Ricky Flint's office under the watchful eyes of the Master at Arms, Andy Reid, and was told in no uncertain terms to keep my willy under control! I pleaded innocence, because it was in the mens shower and she was on top! She shouldn't have been there! After that I think sex was legalised in the work place. We stayed together until I left the ship and she eventually moved from London to join me in Cape Town... but that's another story.

On a visit to Rio de Janeiro, Jimmy Ritchie and his wife Pat joined the ship. I had phoned Jimmy during my last few Nedbank days to ask for advice on my new career at sea, and he was ever so helpful. Over 500 passengers on deck as we sailed out of Cape Town, I scanned the new lot to try and find a magician that I had never met or seen before. I walked up to a couple posing for the ships photographer at the deck rail with Table Mountain in the back ground, and said, "*You must be Jim?*" "*You must be Robin,*" he replied, and off we went to Harry's for a drink. During the voyage I learned so much from Jimmy and it's sad to think that youngsters growing up in the magic world would only hear about the likes of some of our great magicians - like Jimmy Ritchie and Graham Kirk to mention but a couple. Guy's that I had learned so much from.

I performed the Houdini underwater escape during a carnival evening before arrival in Rio. Andy Reid, the Master at Arms, chained me up and hurled me into the pool... every body was there to cheer me on, except Tom & Terry... they were too nervous to watch. The next day in Rio we all toured about and Liz and I nearly got taken for a ride by the taxi driver. *(no pun intended)* He was already taking us back to the ship but hadn't yet taken us to Sugar Loaf. After much arguing he added on about $50 to our yet unpaid bill (Don't pay the ferryman until he gets you to the other side) and turned the yellow Beetle around. We went up the cable car and at the top bought new shirts. When we came down he didn't recognise us and we took a different cab back to the ship and saved a fortune! But we came back to sad news.

Mike Scott, of Scott Free fame, was the trumpeter and bass player for the Peter Adams Sound. They played in the main lounge and backed all the cabaret artists and the Carlo Spetto Dancers. He taught me how to play the bass guitar for *Cherry pink and apple blossom* - he had to play the trumpet for that one. As Tom said it was adding another string to my bow. The passengers loved it when *Mr. Magish* got onto the bandstand... Peter Adams would turn a deaf ear to any bum notes I played... he was a fellow that could read music from 40 yards!

The first day in Rio, Mike was looking the wrong way while crossing a road, and was hit by an oncoming bus. The traffic travels on the wrong side as to what we are used to. Mike was hospitalised and died before the ship left. I was a pallbearer at his funeral in Rio and on our return to the harbour, a British Navy vessel had their flag at half mast for Mike and fired a salute. Mike was buried in the British Cemetery... it was a very sad ship that sailed out of Rio that evening.

Joanne Pezarro and Frank had joined the ship in Rio, to help with extra entertainment for the passengers going back to Cape Town. As it was a return voyage, most of the passengers were the same and we didn't repeat shows. Always something new! It's a bit difficult trying to remember what tricks and jokes you've said and done. I used to have a chart on my cabin bulk-head (wall) that I would fill in after every show... what I wore, what I did, what jokes, etc. It's extremely useful, especially during short voyages when you're on stage thinking to yourself... *I'm sure I did this trick or joke last night!* Jo and Frank were travelling with their adopted baby girl - they had been trying for their own for ages and after they disembarked in Cape Town, they sent a message to the ship saying that Jo was pregnant... Aaaaaarrgh - the ribbon of moonlight... the smell of the life-boats?

"*Be nice,*" Johnny Noble always used to say. One morning in the Seychelles while waiting for Liz to finish in the salon, we noticed a new face wandering about the deck. He had been given a visitors card from Andy Reid and was looking over the ship. Tony Starke, the ships drummer and my official cabin mate for the ships records, and I offered to show him around the ship. After the guided tour we bought him a drink in the bar and he said that he would return the favour tomorrow... *"Bring about ten of your mates, swimming gear and drinks."* Eight o'clock the next morning we stood on the dock waiting for the New Zealander. *Toot - Toot...* not a car or bus in sight, but behind us was this huge yacht, *"Catch the line Rob and bring your friends aboard!"*

We spent a wonderful day with the Kiwi... he was a pro skipper and was delivering the yacht back to France to its owner. We toured the Seychelles that day and their tourism slogan is *Unique by a Thousand Islands... I think we* saw them all and they're not wrong... *Be nice!*

After Cape Town we were heading for Southampton. I was going to meet Liz's family for the first time. She wound them up by saying that she was dating a guy from Africa. The relief on their faces when a white guy stood at the door was quite a picture. We had a lovely belated Christmas dinner and then a cab took us all the way from Harrow on the Hill, back to Southampton... to our next adventure.

Sailing out of Southampton there was a bit of a farewell party going on in Harry's New York Bar with loads of Dom Perignon. When we were out at sea one of the party asked to see John Dimmock, the chief purser. He didn't have a cabin... he was just a visitor on board to come and see his mates off. When queried as to why he didn't obey the *"all visitors please proceed ashore, the ship is under sailing orders"* announcement he replied in the most posh accent, *"I hadn't finished my champagne!"*

He was like a walking advert for the Astor... clothes from the boutique, etc. He had to use the ships satellite phone (no cell phones in those days) to call his wife to fly down to Las Palmas to bring his passport. Apparently he had a little home on the Thames and a Villa in Las Palmas. He became known as the Stowaway for his short little trip, and was a great gambler with Tom and I during the Tote Auction. He loved the game, and Tom & I introduced it to every ship we sailed on after the Astor... to TFC and Starlight ships. (CJ "*Gooney*" Foggit became addicted, I think, to the laughter)

Sailing out of Las Palmas, we had said our goodbyes to the Stowaway. Standing in the Lido Bar listening to *Erezand* (Erez and John Ford) belting out *I just called to say I love you,* Tom went up onto the band stand and just joined in on the piano. Afterwards I asked Tom how he does that. Being a bit of a muso, I know you need to know what key they're in, besides having to know all the chords. He said it was quite simple really... as he gets on to the bandstand he says to John out of the corner of his mouth, "*What key are you in, John?*" Typical Tom!

The same evening Beulah Jacobs, our ballroom dancing teacher and old family friend from school days, spotted a new face at the bar. He was about six foot seven with a huge dagger down his belt. Beulah pulled the dagger out from his trousers and threw it behind the bar. Andy Reid was there in a flash... we had a real stowaway!! He had escaped from the French Foreign Legion and this was his second attempt to get away and into South Africa. The news was in the Cape Argus long before we even crossed the Equator.

How did that happen? Allan Simmonds (of bridge and bowls Argus fame) was on board. His memory is just unbelievable! He had recognised the stowaway's name from a previous attempt to get into our country and survived in the cargo hold of a Jumbo or something desperate. The Stowaway was given a place in the hospital to sleep until they could transfer him onto a north bound container ship, The Helderberg. Two ships passing in the night... not an easy task *(as some of you might know if you lived on A, or C deck!)* Our ship lowered a life boat and all the entertainers were up on the helipad throwing streamers and cheering. We all got into shit for that as well!

Anyway, Andy removed the handcuffs in case of the man falling over the side, and the Stowaway promptly dived into the sea, over the life boat, and swam out into the Atlantic. Commodore Ivan Currie was as quick as lightning. He threw the bridge life-ring over the side and anything else the crew could get their hands on that could float. I thought our streamers would come in handy here... the Atlantic started to take on a sort of rainbow hue of colour.

With the fading light the officers on the bridge activated the searchlights and promptly highlighted two hammer-head sharks! Thank goodness the sharks stayed away... they would never have survived! After two hours in the water our Stowaway finally returned to the Astor and climbed up a rope that somebody had thrown over the side. How fit is that? We were a day late arriving in Cape Town and our Stowaway was welded into a toilet in the hospital for the last few hours if the voyage - in case he jumped over the side in Table Bay and f***ed up our sharks.

It's a wonderful life on the ship. We have the Crossing of the Line Ceremony over the Equator and then we keep our eyes peeled for the Southern Cross as it creeps up over the horizon. After the last show and the disco had finished, we all used to go up onto the helipad and you could only pop the champagne when Table Mountain was in sight. All the gang used to be up there... my cousin

Sharon Withers, Jane Barker, Leonie van de Spuy, Gavin Durrel and Patrick Knight, (if they weren't on the bridge), Larry Jackson, (who taught us *If I were not upon the sea),* Nadia Eckhardt and loads of others.

It was a sad day when we left the Astor - She was eventually sold to the then East Germans, and I thought about what I was going to do next. Tom and Terry had already thought about that. Terry had bought a flight ticket for me to Johannesburg. He had said, *If you can make Tom and I laugh, you'll make anybody laugh.* I was going to be introduced to Jose Broude of the Don Hughes Organisation, our biggest entertainment agency at the time. Tom picked me up from the airport and I stayed at Terry for a while and then with Jimmy (one of the most knowledgeable magicians this country had) I met Jose and between Tom, Terry and Carlo Spetto, I was now up for grabs.

My first gig was Christmas and New Year at the Venda Sun. Terry Lester, Terry Fortune and Tony Starke were all there. It was like a ships reunion. Rod Walker was the GM (currently at Sun International Entertainment) and during a dinner party at his home, his wife Indura had baked a Christmas cake. Rod battled to cut through the icing and after fiddling about with extension cables promptly cut the cake over everybody's plate with a Black & Decker jig saw.

Liz came back to Cape Town. It was lovely to see her again, but unfortunately I was seeing a few girls at the time... you know, the smell of the life boats, etc. She moved in with Froggy (Ruth Frogel, a lovely girl who unfortunately died of cancer long before her time... xxx), Mark (Parrr-Parrr) Rauch - and I was the occasional visitor.

This was also the time when I was introduced to Tim & Neil Akers and would meet their other brother, Pete, a bit later. They were in the motor trade selling tools. They still are today, with Ian Kerr and The Beta and Teng Tool Imports Company. Neil has since moved back to the UK. We've all remained friends; in fact they are now my family because their nephew Michael has married my sister Jennifer. Tim and his wife Mariaan always accommodate me when I visit Cape Town.

Anyway with all the night life, magic and mixing with entertainers, you get to meet loads of interesting people. Strippers/dancers in those days had to wear nipple caps - but not in the dressing room. I don't know how it happened, but one day I caught crabs! Unfortunately, before I realised it I had given it to Liz. She wasted no time in telling everybody, Tim, my Mom, Froggy... the lot! Very embarrassing, to say the least! I had to pay for all the flats laundry, anti-crab stuff and whatever else Liz could think of. I phoned my pharmacist friend, the model and Hellenic soccer player, Paul Danielz. He gave me some potion and the crabs were cured. I've since learned that you can also sprinkle icing sugar on your willy... It doesn't cure it, but it rots their teeth - and you get a decent nights rest!

Years later I was doing a show and golf day for Wool-Tru at the Paarl Country Club and Paul was in the audience. I told the story about the crabs, while on stage during my act, and said that I had not really thanked him because it was a bit embarrassing, etc. But now with the microphone I thanked him in front of all those people. They laughed, but not as loud as Paul and I did... we knew it was true!

Liz has since moved back to London and is running her own hair salon, but she came over to see Froggy when she was ill. We said goodbye to Froggy and she was very brave right to the end and still had a wonderful sense of humour, although she was in constant pain. I drove Liz to the airport and the next day Froggy left us forever, leaving only fond and loving memories…

THE WILD COAST SUN

Carlo Spetto was producing an extravaganza called *Just for Kicks.* He had met me on the Astor. Loads of beautiful girls, including Lee Tyler and Karen Peacock, dancers from the Astor, two male dancers, Waldo Cottle, (who looked like Rambo), Tienus Maree of *Soft Shoes* fame, and four *Stars!* Danny Fisher (great to see him again), Terry Fortune (Astor) Adele First who sounded like Whitney Houston and *Mr.Magish.*

What a great show it was. It was due to run for three months, but it proved so popular that the show was extended and we did four and a half, over 280+ performances. I introduced myself to the stage managers, Allan Michie and the Wee Gerry Dunn, and his reply was *"Shut the f*** up!"* I thought, *"Here is a fellow I'm going to like."* Danny Fisher invited me around on the first weekend during rehearsals and I couldn't make it. I was already invited over to Allen and his wife Lisa for lunch... Danny said*, "How did you manage that?"* I was to see Allan and Lisa and the wee Gerry a lot more over the years in my travels.

I thought The Wild Coast was named after the Coast Line, but I was wrong. It's named after the girls! I have never been involved with so many girls in such a short time in all my life. This is not circumcision... its wear and tear!

As I was sponsored (paid a retainer) by Martell Brandy in those days, after my involvement with SFW and the Sedgewick brothers, I used to hand out a miniature bottle to anybody I used on stage. Needless to say backstage there were loads of miniatures too... all empty. Danny, Waldo, Tienus and I shared a dressing room with the Martells and Terry was a *frequent flier,* and visited our dressing room for a glass of Martell... just to warm the vocal chords! (He would often say that he wouldn't harm a fly... well, not unless it was open!) The girls were next door. Outside their door was an ice bucket and on their way to the stage they would rub their boobs with ice blocks. (It was a topless show) I thought, *That's the job for me!"* The girls took it all in good fun but Danny couldn't believe his eyes when I first started rubbing ice-blocks on their boobs. He used to complain to June (Spoon - his wife) that he couldn't concentrate.

A new girl joined the cast. Tanya was beautiful and we soon got together. I went to share accommodation with Terry Fortune and Tanya moved into my old room at Waldo's 'Rambo Gym.' At a birthday party at Manuela's house, I was introduced to her and she asked me to show her the way home on the staff complex. The next day when she came into our dressing room, topless, and gave me a kiss hello... Danny got the shits again!

Sharing a house with Terry Fortune is another story... man, can he cook. In minutes there would be the most amazing curry for anybody who popped in to enjoy! He enjoyed *smoking* and in case the police ever popped round, he had it hidden in a plastic Cremora jar in the garden tied to a piece of string. (*it went inside, - not on top!*) Every time he wanted a joint, he would reel in the Cremora jar, roll it, and then hurl the jar back into the bushes.

I borrowed Adele's car one day and went off in search of a Christmas tree. I thought Terry was going to cry as in all his travels he had not had time to be in one place long enough to enjoy a simple thing like putting up a Christmas tree.

The dancers spent a lot of time down at Water World with Flynn and Graham t/a Harry. (There were a few Grahams) The dancers used to tan topless and they were always a great draw card. I can't imagine what the people thought driving over the bridge on the Umtamvuna River while Tanya was 'Water-Worlding' me on top of the Spellbound Gambler!

Sunday nights at The Old Ferry restaurant were some of the best nights that the people of Port Edward enjoyed. Dix Foster and Brenda were the hosts and Danny, Dix, Terry and a resident local called Dexter Ronniger and I would entertain with guitars and my banjo until the wee hours of the morning. It was also here I met Graham Sheville, the owner of the hotel in the Natal Midlands - the Argyle Arms Country Inn.

Dix Foster, me and Danny Fisher

On my 27th birthday, we had a marvellous day at Water World. I was an Alan Alda fan and I wondered how he could be so funny in the M*A*S*H. series. It must be the Martini. I finished an entire bottle, took all the dancers off to San Lameer for pre-show drinks, up to the golf club to have drinks with Fred Beaver (the Pro) and then to the shows. We did two shows a night. Between the Martells back stage and drinks at the bar with Danny in between shows… I was fast becoming the Dean Martin of magic. He once remarked, "You're not really drunk if you can lie on the floor without holding on!" That night I resembled that remark. I had to do about 15 minutes on stage in front of the curtains while the back stage crew changed the scenery for the grand finale. During my act the cast stood behind the curtain and sang happy birthday to me and the audience of over 400 joined in.

I did about 40 minutes that night and Allan and Gerry in the wings were just urging me on - I got a standing ovation! The next day Alberto Chiaranda (Chico - the Big Boss) wanted to see me in *my* office. (The Lagoon Bar) He was with another guy and first he wished me happy birthday and asked about my contract and drinking. I said there was a clause but it said *unable to perform.* I said I even got a standing ovation last night. Then he introduced me to the man alongside. He had told Alberto that he had not had a laugh like that in years… anytime I was in Durban I was invited to stay with him. His name was Gottfried Gratzel and he was the GM of the Maharani Hotel. I think I stayed in every suite in that lovely hotel on my Monday nights off. I got the thumbs up from Alberto.

During a strike at the hotel everybody helped with everything. Croupiers even made the beds, and Sol Kerzner was flying in to help sort things out. In the bar before the show the barmaid apologised about the lack of glasses and I decided to help her with the washing up. Alberto came in to collect two glasses and a bottle of Jack Daniels for Sol. He looked over without batting an eyelid and reprimanded me for not wearing a name badge. With all the staff problems he was experiencing, running a huge resort like he was, the next day in my dressing room was an envelope addressed to me. Inside was a Sun International name badge - *Mr. Magish* He still found time to think of a little thing like that. What a guy!

JOKERS WILD

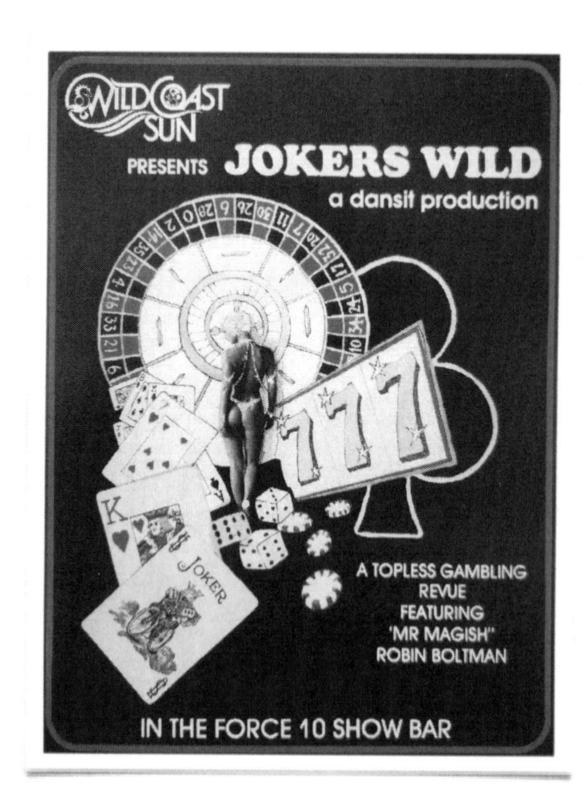

Towards the end of *Just for Kicks,* a new show was in the pipeline for the Force 10 Show Bar. It was called *Jokers Wild.* The show had a gambling theme and was based largely around me. I even had a small dancing roll - after 80 odd performances, Danny said the show wasn't as funny anymore - I had finally got the steps right! (I've got the rhythm of a jar of Marmite). We did over 300 performances and I received some wonderful write ups.

I met the new cast other than Mich (Michelle was in *Just for Kicks*) their names were fascinating - Dee, (Deanne) Tee, (Theresa) Gee, (Guiliam), ZiZi (real name), Den (Denise Palm - the producer), and Roo - Andrew Burns my favourite sound man. (Dee and Andy - Roo are married and living in Sun City)

I was the only one with a real name that I could remember! It took time but I remembered… I'm actually writing this book by memory.

Next door to the small theatre was a wonderful restaurant called The Commodore. Betsy Van Rooyen used to play the piano and I would spend every break between shows listening to Betsy. She had her own Vodka and Tequila in her handbag, a novel and her smokes. Betsy would read, drink, smoke and converse with the patrons seated around the piano and not miss a note. We used to play trivial pursuit on the piano and after one session after she played *Polka dots and Moonbeams,* I sang along, she said, "How old are you really?" Tom Hine used to "dep" (stand-in) for Betsy every August when she took her annual leave.

I left the Wild Coast after 11 months, loads of friends later and a willy that needed a break. The GM from Tha'Banchu Sun had seen my act at the Wild Coast Sun and knew I could speak Afrikaans. I was now going to entertain the Free State Fraternity at the Diamond Lill's show bar.

THA'BANCHU SUN
"Black Mountain"`

The Big Sky Country... Diamond Lill's... I called it Dreary Lill's. Not a lot happens there. I met some lovely people there and Duncan, Joy, Willy and Kim Dundas, with Mike Flowers. They were the fun lot that weren't in the show. All from Cape Town and Mike was a Geordie, like Tom, from Newcastle. Kim looked like Gabby Sabattinni and I felt like I had just won Wimbledon... Tim Akers (the tool man) came to stay during Christmas and New Year - my bar bill was a fortune! We got paid once a week and I went to collect my salary on Friday and *paid in* about R500-00 Rand! Thank goodness he left after 10 days. Enda Mullin (GM) saw us in the pool one morning with beers and the water from the fountain cascading over our shoulders and said, *"I can't believe I'm paying you for this?"*

Tim, being the English gentleman that he is, sorted out a lot of the English croupiers problems... a bit like Dr. Tim at your cervix! By the time Tim went home I had to deal with a few broken hearts and a few other parts of the female anatomy! Sidney Hillman and Hillary Dollman (army) also came to visit, and in years to come, Sid must have visited every venue I have ever worked in. For their friendship and loyalty to supporting any show I did, I even got them on a sea cruise in the years to come.

In the show produced by Carlo Spetto, the other specialty act was the glamorous Sophia Foster. She and Craig Steyn were to eventually sail with me on their honeymoon on the Achille Lauro. One evening before the final bows in the show, I had changed costume and while washing my hands, I squeezed white liquid soap into my left hand. That gave me an idea... when we took our bows with the girls at the end of the show we would hold hands and do a final bow. When they announced Robin Boltman, on I strolled, took my bow and pretended to sneeze into my left hand which was filled with the liquid soap - grabbed Sophia's hand and she screamed and shook the white stuff off... over the audience and the dancers - pandemonium and laughter! Sophia mentioned this during her episode on *Late Nite Live*.

Jo-Anne Pezarro was there for the New Years' Eve bash - I hadn't seen her or Frank since the Astor and Rio - what a pleasure working with a pro like Jo - she knows how to handle a microphone... what a voice!

I stayed there for a while and the staff pub decided to have a party for their first birthday. They invited a guy called Tony de King to come down to entertain the staff. He had worked the hotel before and the staff loved him and his show. Tony was one of the first entertainers to entertain the troops during the Falklands War and now he was here with his lovely wife, Nicola. I had been booked to do a stint at the Jan Smuts Holiday Inn where Tony was currently working after I finished at Tha'Banchu Sun. Tony was brilliant and he said I would crack it at the Runway Bar. "*I couldn't follow that,*" was all I could think about. He was to be my best man at my wedding in years to come, and I'm God father to their children, Kimberley and Jason. What an act I had to follow!

When I eventually left Tha'Banchu Sun and moved to Jo'burg, I invited Tony down to do my farewell party in the staff pub. An English Pub-Night with mushy peas, fish (long time no sea) and chips and a shit load of drinks. Tony and I entertained the staff into the wee hours of the morning. I had bought polystyrene hats and painted each one with a Union Jack, all that is for except for one - Enda Mullin is from Ireland - he had a green one!

Tony and I missed the plane the next day we and had to take a later flight. Taxing down the runway on take-off, Tony started shouting... *"We're going to die, we're going to die!"* I hope he never grows up.

Robin Boltman and Tony de King.

THE RUNWAY BAR

My first afternoon at the Runway Bar, I said, "Good afternoon, ladies and gentleman." A voice from the dimly lit audience replied, "F*** off, bring on the tits!" I thought this will be an interesting gig! (My mom said there would be days like this - she didn't say it would be today!) It was a strip joint and all the big names had already worked there; Joe Parker, Tony and Barry to name but a few. (No, not stripping... we were the Comics, MC, etc.) It was here I was introduced to TV people and agents like Mike Bourne that got me onto TV. I did *Take a Break*, *Late Night Live* and a few others.

My mate Sidney was there too, for the filming of *Late Night Live,* with Tony Sanderson. He was never seen on camera... I was having an affair with a beautiful girl and Sid sat next to her. I told the camera men not to go near her. (Names are changed to protect the guilty) The beautiful Michelle!

Colin Senekal and some of the regulars at *The Runway* are still my mates to this day. Barry Hilton phoned from England to wish me well in his old stamping grounds and after two or three days the locals handed me a present - it was a microphone. Ron "Spitfire" Woods used to call me Robin *one trick* Boltman. The boys in the pub didn't want to see magic... they were there for the birds stripping on stage.

M-net had just been introduced to the country and not many people had it at home. The Holiday Inn did and one day Colin Senekal and Felicity (still friends to this day) Ron, Nick, Andy, Cyril and a few others asked if they could come round to watch the boxing in my room? Not a problem! I didn't know what time it was on. At three in the morning (when most people can't even argue… let alone fight) the guys arrived with beers and about eight people got into the two double beds and we settled down to watch the fight. The real fight only started when the wives started phoning at about eight in the morning and everybody was rat faced!

I was booked to do the lunch hour and the evening show Monday to Friday. The ever popular band, Ballyhoo, played in the evening and we all lived down the same corridor. I was in room 33, right next to the pool and the pump. By the time I left there, I new all of the migratory patterns of that f***ing Kreepy-Krawly by heart.

Every year my New Year resolution was to learn to play a new musical instrument. This year it was a trumpet. I just couldn't get the hang of it; maybe I had bought a dud! Tom Hine came to visit and I asked him to check it out for me. Tom could play the trumpet but his forte was really the piano. Anyway he blew it and Tom turned into Satchmo - *just like that!*

I practiced some more and one day there was a knock on my door. The entire Ballyhoo band was there with a present - it was a trumpet mute!" *Stick that in the front of that f***ing trumpet,"* said Fergie! Maybe I'll try again if I ever get stuck on a deserted island! (This is quite possible if you read on.)

From *The Runway* I went to Gold Reef City and worked with some old friends - Ronnie (*lend me ten bucks)* Hendrickse was a maestro on the piano and our neighbour's, Des Lee and the band *Cinema.* The Consolidated Bar had Can-Can girls… and they *did-did!*

The very pretty barmaid, Verity (who was often on the back page of the Sunday Times and other girlie magazines) Linda and Ronnie and some of the girls and I would go out on Sunday nights to *Springsteen's* - Erez (Astor) and his family owned it. Erez and I both agreed that we had to get back to sea!

I moved out of Hilbrow and moved in with an old mate who booked me for Southern Suns and Holiday Inn, Clive Levy. Between Clive, Jane (now his wife), Heather von Reuben (a great friend and Clive's boss) and any stray I could find, Blairgowrie was never going to be the same again.

John Wainwright and Mandy lived in the same road. John was an amazing talent and he used to perform at *Hunters* in Randburg. Jane was an air hostess and she would often bring some of her hostess friend's home. Good girl! I convinced Marisa to stay one evening, "*You can't drive all that way home!*" (15 km) She was a noisy girl, and Clive didn't get any sleep that night. We dated for a long time but then she moved to Australia to work for Qantas.

I was booked for a week or two in the Gaborone Sun in Botswana. The way out in those days to avoid drama and paper work was just to say you were on holiday. Remember this was the late 80's. Purpose of visit the form stated, and I had filled in "Holiday". The customs lady said, "Holiday? Are you sure? You are Robin Boltman?" Yes, of course this is me! She pointed behind me and I turned and looked up to where she was pointing... there was an enormous advertising sign - Gaborone Sun presents... Cabaret Magic and Comedy with Robin Boltman! I was taken to the police station and Gab Sun had to sort things out. My passport was taken every day to the station to be stamped!

The Customs lady and her husband came to the show one evening and said she was pleased they had sorted things out because they had had a wonderful evening! I should have used my Hout Bay Passport!

Nadia Eckhardt, Jannie Cloete, Tom and a few others from the Astor were employed by Allan Foggitt and TFC Cruises. The Achille Lauro was starting the season and they decided that they wanted me on board. I left my worldly possessions (golf clubs and my Datsun '*girl - repellent'* Pulsar) in Clive's capable hands. He was an estate agent and he used to use my car when he was showing houses. People used to feel sorry for him in this old f***ed out car, and buy the house! It's a ploy that I think the *mucky bastard* still uses to this day! Anyway I was flying home to Cape Town to join what was to become home for me for more than four and a half years of my life... the cruise liner, world famous after the High-jacking in 1985 - The MV Achille Lauro.

MV ACHILLE LAURO

A beautiful ship weighing in at 24000 tons and I could smell the life boats already. I had to share a cabin with a TFC host, Tony Karim, and together we worked out a code for our cabin door. Prestick (Blue-tack UK) on the left corner - safe to come in - Prestick on the right corner - find your own lifeboat!

We had some wonderful entertainers on board including Michael *yebo gogo* Depinna. He was the Entertainment Director, and with some superb entertainers at his disposal, did a wonderful job. Jean Dell & husband, Dennis Smith - remember Springbok Radio? John, a magician from England, used to get the shits when I was on stage. I used to use English and Afrikaans in my act and had the audience in an uproar… *What did he say? What did he do?* He couldn't understand it.

Martin Clifford, probably the best puppeteer in the world, had his Biggest Little Family on board. Martin had spent more than 25 Christmas', New Years and birthdays at sea, probably much more by now. Skip Cole was a comic from Bradford who started out as a ships gorilla (ship photographers would take photos of passengers with the gorilla). He progressed from gorilla to DJ and eventually as a comic by copying the best comedians that he had met and seen on the various cruises. I used to tease him before he went on stage. He would write things down on his hand so he wouldn't forget his lines or his gags and just before he would go on stage, I would say, *"You're going to die again!"* We became great friends. Years later he and his folks attended my wedding.

After a few drinks one evening after the shows, he jumped on my back and shouted, "*Comedy Staff on Tour!*" Off I ran to the top lounge where Tom was playing and ran right into Captain Giuseppe Orsi. "*Road block, road block*," shouted the growth on my back, and off we went like naughty school kids running down the promenade deck. It took a long time before the laughter died down.

Skip wrote a song one day and he thought it was very funny. It was even funnier watching him trying to communicate with the Italian Band in the front lounge as to how the tune should sound. Christmas Eve, Mike separated the shows - black tie affair in the top lounge, and casual in the Scarrabeo Lounge, where Erez, John (Astor) and Peter Ball were playing. Skip, armed with the words of his song written on the back of his hand, a few doubles down his throat for courage, was in the top lounge, ready to conquer the passengers in the Arrazzi Lounge before the Christmas Bells tolled. Bearing in mind, during rough passages, the ship puts out little tin-foil packets in case you feel sea-sick. The song went something like this…

I've thrown up in the foyer
I've thrown up in the gym
Now they've given me this little bag
To keep the big bits in!
(Chorus in harmony with the Italian band who incidentally didn't know what they were singing…)
Throw it up, throw it up
Throw it up, up, up… Blaaaaah!

I suddenly saw a long faced Skip in the lounge where I was working. Our party was in full swing and I asked Skip how his song went. "*Michael and the ships minister told me to f**k off!*" he said. I don't think he sang it again.

I had to take a few days off because of a previous booking at San Lameer. It was the Miss Teen beauty pageant… I didn't want to miss that. I phoned Dave *Kiwi* O'Connor, the Bing Crosby of Magic, and Dave worked on board for me. Lisa King of Bantry Bay in the Cape was declared the winner and my old mates from Gold Reef, the band Cinema, played their hit single, *My kind of girl,* and all the girls and parents had a fantastic time… especially the top three anyway!

I flew back to Cape Town to rejoin the Achille and Dave looked like he hadn't slept for the three days. He had loved every minute of it and the team loved him. He loved it even more when I gave his return airline ticket in business class! Its great to have friends you can count on.

On the northbound passage we had twelve of the most beautiful girls for the old Scope magazines beauty finals and Dale, who had just won the title of *Mr. Sandton*. Don Martin was on for this function and he and I had sailed on the Astor, and all the different ships to date. Chris Cummings from the Astor was also on board and I know that he didn't get much rest. What a lovely cruise that turned out to be. Melanie Walker, Patricia Lewis and the wonderful Alex Klootwyk were just some of the beauties that are still around. A Durban girl, Karen Eisella, did the bra *Stikini* ad and later became a model in the USA. They were so polite to everyone... they didn't know who the judges were!

They also tanned topless with our dancers on the top deck and one of the girls asked why I didn't try to hit on them like most of the other guy's had? I replied saying that I was after that one over there; one of our dancers... Gail turned red and I removed the Prestick off my door and put it on hers.

Gail was lovely and I spent quite some time with her in England. Liz came to fetch me from Southampton and Gail came to stay for a few days. Here I was in a flat in Hammersmith with an ex and a current girlfriend. Oh, the smell of the life boats! The first three days were the hardest!

After the first Achille season I went and stayed with Skip and his folks in Bradford for a few days. Great fun was had in his local *Swing Gate* pub. He introduced me to all his mates and of course the magic was fun for the locals. One night we were invited to a lock-in party after closing hours. I seem to recall Nan and Fred, Skips folks, trying to pull us out of the front hedge in the early hours of the morning. When we left England to fly home, I invited Skip to visit anytime he liked. He arrived a week later in Johannesburg and stayed for about four months! Thank the stars that Clive and Jane liked him. I took Skip to all my shows, TV gigs and to Sun City a few times. Chris Cummings (Astor) started the Sun City Express. I was the MC on the *Express* that took about 5 hours to get from Johannesburg to Sun City. Skip went home and I went to see Allan Foggitt to make sure he knew that I was available for the next Achille season. I handed the Datsun *girl repellent* back to Clive, (to help sell his houses) and off we went to JHB International.

All the *shipping family* were there - Tom, Terry Lester, Terry Fortune, Carlo Spetto, Michael, Nadia, Jareen, Geraldine Massyn, and a few new faces. There was a beautiful girl at the airport, in the Alitalia queue with her mom, and I hoped they were also going to be on board. Down in the pub, once the passport formalities had been completed, we started on a few drinks to get into the mood. Carlo is not a good flier and he has quite a few J&B's. Terry Lester's blood group was J&B positive and Tom and I had the last few Castles before we had to convert to Perroni - Italian beer.

I sat next to a Greek Goddess and we chatted into the night with loads of drinks coming our way. During the night she lifted the arm rest and snuggled under the blanket… I thought perhaps she had lost her passport down the front of my trousers or maybe was in search of my money belt! Anyway, everybody else was asleep and I let her continue with her search - uninterrupted! The life-boats were getting closer…

In Genoa Carry-Lee Cunningham, a dancer from the Wild Coast days was on board as one of the Carlo Spetto Dancers. With her was the girl from the JHB queue, without her mom. Lorraine Betts was our Cruise Director, and wasted no time in sorting out cabins and staff. My new cabin mate was Ivor and he was already in love with Ilsa, a dizzy blonde dancer who wanted to remain a virgin for as long as possible. Poor Ivor - his mates, Neil Wesselo and Rob Wheatley from Radio 702, were also doing well in the spade work department… carrying bags for the girls, etc. Martin Clifford and the “Biggest little family” were there and another Magician called Julian Russell, from Yorkshire, who instantly became one of the team. After a few drinks, his party trick was to dive off a top bunk into the *tear of an ant!* Johnnie Jackets (Baatjies) was the leader of the band and we had an English Band in the main Arrazzi Lounge. Moss and Tracy played the cocktail set and the late evening session in the Sorrento Bar where our resident barman, Giorgio, was to become our best mate and new dancing instructors, Van and Monica.

We had a new Captain on this voyage - Captain Gerardo de Rosa. The previous Captain, Orsi of the *road block* fame, used to have no entry signs all over the place and I queried if we can organise bridge visits for the passengers. He wanted to know, if he ever visited me in my home, if there was a room he would not be allowed in? "My ship is open to al, the passengers can go where they like!" He used to take over from Orsi and the first thing he used to do was throw all Orsi's - *Do not enter* and *crew only* signs over the side. What a great Captain... more about this hero later...

Captain Gerardo de Rosa

Tom wasted no time in getting the Tote Auction up and running. The two of us had a helper this time, the always smiling Jane Barker from the Astor. Jane also did the accounting for TFC and she handled the money side of the auction, Tom was the auctioneer and I was there to help him remember the passenger's names. To Tom everybody was *my little love* and *my little pet!*

The gorgeous girl from the queue in the airport kept walking through the lounge where we did the auction and I used to lose concentration. Her name was Carol and boy could she dance. She also did some contortion type of dancing during her solo act and she could get her body into some wonderful positions... which would come in handy later when she saw the size of my cabin! Ivor was still panting after Ilsa, Neil and Geraldine got together and Rob was sorted. After Tropical night out on the pool deck Carol asked me to dance after she had done her first solo performance. All the passengers were congratulating her on a wonderful performance and she was in heaven. Carry-Lee looked over towards us and said, "*It's about bloody time!*" (I was sometimes slow in picking up signals) I took her on a tour of the life-boats and we stayed together for a long time, even after the ship. She lived in Blairgowrie, two streets away from Clive and my girl repellent Datsun!

There are so many regular passengers during the season, but not nearly as many as during a north or south bound trip. Pat & George Gardner, Stanley, Chris & Dede, Allan East, Eric & Andre Jacobs, Pat and Lionel Roche, Geoff & Mary Hargreaves, Brian Hineson and Peggy Grinaker, to name but a few. If I have left anybody out, I do apologise, but I'm doing this all from memory. Terry and Martin got involved with the auction and it just got funnier as the days progressed. Like I mentioned earlier, it is a gambling game based on the ships mileage. Passengers bought Nom de Plumes (false names under which to bid) and they all had some funny connotations. All that is with the exception of Conrad Nissan - he was always a Peregrine or a Daffodil or some squeaky clean thing.

There were small, big or tired willies, *Le Gover* (or better know as Leg Over), Russian ballerina's - *Nora Titsov,* and her partner, *Ivor*

Nakerov. (They could only dance in a clock-wise direction for obvious reasons) Tom had a field day with some of these. The names would be written up next to a mileage and if a passenger wanted to buy it they would have to bid using real dollars. Tom would bellow, *"I'm selling this little Willy - who wants a little Willy? I've already got one, as my ex wife used to say, "Such a big fuss over such a little thing!" Anyway this Willy is being sold Voetstoets... As it stands!"* The passengers loved him and the Italian casino staff would watch in fascination and horror, as money changed hands and they had no idea what was going on and they weren't getting any of it!

During the first few days of the cruise, passengers used to wander around aimlessly for hours getting lost. The Achille used to be a first and second class ship, and all the stairs and lifts went to places where you didn't want to be. In the lounge where we held the Auction was a lift and a stairwell with a map of the ship displayed prominently on the side. A lady was studying the chart (which had a little dot saying *Landmark,* of all things, instead of *you are here.*) She turned and asked me, *"Do these stairs go up or down?"* Well, the Auction came to halt and it was quite sometime before we all stopped laughing. (Now you know where I got the title for the book)

There was $10 up for the highest bid of the morning. Tom used to frequently remind the passengers and Jane would point to the note which was stuck next to a switch. One day she accidentally knocked the switch and the fire doors flung her into the Sorrento Lounge! On very rare occasions if Tom wasn't on board, for some reason or other, (he was also a cocktail pianist on land and sometimes could only join us at various times during the voyage), I would do the Auction. Selling low and high field, the captain's estimate, we had stories for every one of those 31 miles on the chart. One morning we seemed to be going very slowly (*on both sides of the ship*) and I was about to sell the lowest mile - Low Field. I told the assembled team how slow we were traveling - *Was the sea and the wind against us?* Or were we just carrying a lot of extra weight on board? Just then the lift opened and the largest, fattest lady on board popped out! Jane shot into the Sorrento Lounge with the same speed that the snot shot out of my nose! Bedlam reigned! Stanley was on the floor and George and Pat decided to crawl under the chairs in the pretence of helping Stanley to his feet. The poor lady didn't know what we were all laughing about. The auction finished a lot later that day.

Tom was free of his previous gig by the time the ship was halfway down the West coast of Africa, and during a party in Pat & George's cabin on the Lido deck, I happened to mention that we must have a drink for Tom on the 6th Dec for his birthday. George and a few passengers had a better idea. I used the ships satellite telephone and organised for Tom to be flown to Walvis Bay to join us for his birthday. Between George and his cheque book, TFC and Nadia drove Tom off to the airport from Cape Town. He joined us in time for his birthday. That is how much the passengers loved him.

AUSTRALIA

After our South African season, the ship was sailing to Australia for a taste of the South Pacific. Lorraine had officially re-arranged the cabins into couples and Carol and her contortions were now a permanent, pliable fixture in my cabin. Ivor was still barking up the wrong tree and I tried to explain to Ilsa that Virginity was not all it's cracked up to be, "The longer you keep it, the less its worth!"

One afternoon Captain de Rosa dropped in during one of our afternoon gatherings in the Sorrento Lounge. We were just chatting about ideas and he noticed Carol was learning to roll a coin across her fingers - a magicians thing. Capt de Rosa said, "I see you are learning some magic Carol. What are you learning from Carol, Robin?" I put my drink down and folded myself up and backward rolled over the back of the couch and into the pot-plants! The laughter took a while to die down.

We sailed over to Fremantle via Mauritius and lost an hour each night with the time change. Onto Melbourne, and finally into one of the most beautiful bays in the world, Sydney. The weather was perfect and in Melbourne I had used a cell phone (1989) for the first time. I had called my old Cape Town neighbour's, Steve (my Gynecologist friend) and Cara, his wife. Another call I made was to my old air-hostess, Marissa. They were all on the dockside and it was a very difficult day out trying to keep Carol and Marissa happy. I survived, but only just. Tom recited a little ditty for me when I arrived back and it went along these lines…

The Duke of York had a hundred wives,
And he serenaded them daily.
But whats the good of a hundred wives,
When you've only got one ukulele?

At the beginning of one voyage, a passenger complained about the view out of her cabin. I said*, "If it's a life boat, unfortunately it's got to stay there!"* On checking her cabin number I realised she wasn't on the boat deck, so I accompanied her up to her cabin and saw this huge grey metal obstruction outside her porthole window and said to her... *"It's a crane and it belongs to the docks... when we sail out it will stay there!"*

On the way back, Julian Russell and I had to entertain a group of travel agents. They came on board to help promote the ship for the next season. During his act he told an Irish joke, and an Irish passenger that was traveling back to England with us, took offence. Lorraine gave me a snotty letter that he had written to her and she asked me to sort it out. I'm not stupid - I gave it to Tom! The ever diplomatic Father of the Seven Seas wrote the most amazing letter to the passenger - *As an uninvited guest at a private party, threats to staff,* etc. suggested that he conforms to the rules of the ship and hopes he will be able to complete the following 40 odd days at sea without being disembarked at one of the ports of call. The poor Irishman made the trip but had to endure Terry on stage with gags like, *"There were these two Peruvians... Patrick and Murphy...!"* I don't think he attended too many shows!

We picked up two lovely girls in Mauritius. They had flown over for a holiday and were now sailing home. The teller from Nedbank in New Germany had a body like a page three model and Sinatra blue eyes. I was on the pool deck preparing questions for trivia that evening when she suddenly dived into the pool. She came out up the steps and I said that is the most amazing sight I've seen in a long time, so she dived in and did it again! Wow… a few nights later before we docked in Durban, we were leaving the disco; she called me over and asked if that girl ever left my side. I could escape for a while, I told her and asked, *"What's your cabin number?" "Foyer deck 36"* was the reply… *"See you in 10 minutes!"* A bit of *Tag and Release - Off Shore Drilling - the moonlight, those life-boats!*

We eventually arrived back in Southampton and stayed with Liz again. I took Carol up to see Skip and the family and then she went back to South Africa.

For the next season, 1990/91, Allan Foggitt asked me to be Cruise Director and Entertainment Director. I was a busy *buoy* that season. I started early… even before the ship sailed out of Genoa I had hooked onto a new dancer called Joanne. After a major reunion at a pub in the town, we introduced some of the new crew to some of the old. Sean Anderson was going to become another regular. Sammy and Sharon Hartman, from Cape Town, Johnnie *Jackets* Baadjies, Zayne Adams and the band, along with Tony (Astor) Starke. We went back to the ship and finished the bottles that Julian had bought ashore. When he decided to dive off the top bunk, (into the tear-duct of an ant) I decided I would have to do some diving of my own. Jo and I stayed together for the whole voyage. Graham Sheville (Wild Coast) always joined me during Christmas or New Year and shared my cabin. He loved watching Jo stroll naked to the shower while he pretended to be asleep! *Mucky Bastard!*

Captain de Rosa was on board again and ex captain Orsi's *No Entry* and "*F*** Off* " signs flew over the side. Captain de Rosa was the hero during the high-jacking of the Achille Lauro in 1985. The movie *Voyage of Terror* was filmed on board and starred Burt Lancaster as Leon Klinghoffer, the wheelchair bound passenger that was shot. Neville Williamson (Magician pal) and Diane Head (from TFC Head Office) were on board during the high-jacking. We had the movie on board and after the movie we used to do a questions and answers session with the Captain. I used to have to sit by his side during these sessions to help him understand some of the marvellous accents our colourful country has to offer. He was a very modest man but his butler, Lorenzo, told me some of the frightening moments he had to endure. A terrorist would want to shoot a passenger to get the authorities to realize that they meant business, and the Captain would calmly pull the barrel to his own chest and say, "*Make head lines - get your demands met - shoot the Captain - but don't harm the passengers!*" He stayed on his feet for the duration of the Hi-jacking... what a man!

Eventually after all the drama was resolved, the Italian Prime Minister phoned the Captain at sea and warned him that an organisation claimed that they had planted a bomb on board. The Captain knew that only water and fuel were loaded on during the last port of call - and a new casino! He didn't know how much time he had, but he didn't take any chances. He mustered all his crew to meet him in the casino via the bridge PA system and they followed his example as he promptly threw every table, chair and all the one-armed bandits over the side. When the casino was clear he wiped his brow and announced to the wide eyed crew, "*No more bomb!*" During bridge and engine room visits with passengers, I always used to point out the old bullet holes.

He loved setting off fireworks during Tropical Nights on deck. He would stand over the mortar type tubes that he had the engineers make with a cigarette in his mouth and send rockets and flares into the night sky. On my birthday before Tropical night we had him speaking like Donald Duck from the Helium gas with which we filled the balloons. Then he asked me to accompany him to his cabin and knocked on his own door. He used to have a regular visitor called Celia and only a few of us knew about it. "*Avanti*" and in we went. They handed me a birthday present and we had a swift glass of champagne and he said, "*Best you go now, it's not good to see a Captain with tears in his eyes!*"

In Durban one day I had a surprise visit from my old mate, Neville Williamson. He was on his way to the airport after doing a gig in the area. Remember, Nevill was on board during the high-jacking. He asked to come on board and I took him up to the Captains office on the bridge. I knocked on his door and he said he was busy with the ships agents. I told him this was more important, "A*vanti, Robin*" and when he saw Neville in tow, the reunion was something to behold. The ships agents had to wait. I'm not sure if Neville made his flight on time that day.

When we left Durban to head off for our Australian season, some new guys joined the ship. They were an Australian band under the leadership of Ken Tweddle (an Englishman Geordie) and a new Cruise Director called Johnnie Smythe (of *Winchester Cathedral* fame). I would become assistant cruise director in Ozzie.

Chris and Dede sailed over with us and we had some memorable times in their cabin. Eight cruise staff & entertainers piled into their cabin one night and played Obbly-Dobbly (a tongue twister game resulting in your face being covered in black Oobly-Dooblies from a burnt wine cork). The next day when Tom got up, he thought he had contracted African Measles!

It wasn't a happy ship that sailed out of Durban; loads of the passengers were immigrating to Australia (remember it's1989/90). Big John Morton had being playing for the Natal Sharks and they had won their first Currie Cup in years. He was happy, as was *Cheesa Labuschagne* (Stuart Brown) of the TV show *Villagers* fame. On our first night out after the show, I was sitting with Geraldine, Neil and Sean in the Lido bar. It was late and not normal for the tannoy to be used. *Ships doctor please phone 320* (Bridge). Two minutes later it went again and the four of us shot down to the hospital on Foyer Deck Forward. There was a family with a little blue baby battling to breathe. The doctor arrived and produced a nebuliser with a South African plug on it. I dismantled some of the bands gear and attached the Italian plug in what seemed like hours. The wife was screaming and Geraldine was doing her best to calm her down and I was armed with a cocktail knife trying to be McGyver! We got it going and we sat with dad and the kids while mommy was inside with the baby. She joined us at about three in the morning with a pink baby and we all cried. We all cried again when they said goodbye and disembarked in Fremantle.

A Wild Coast, Water World, Colin was on board for the crossing, and unfortunately he and a lot of our team, Jo and the dancers, Moss & Tracy included, had to leave and make way for Ozzie Staff and entertainers. The new lot was a great bunch and I met some wonderful entertainers and passengers. We had to learn a song that Liza Minnelli's ex husband; Peter Allen wrote… *I still call Australia home.* We did that song at the end of every voyage, with all the entertainers on stage and there was never a dry eye in the house.

A passenger died before we sailed out of Fremantle and Mario (security) and I were the first to answer the screams from his wife. We tried everything to revive him, but to no avail. I had to go ashore and find his family and his daughter said he had lost his first wife at sea, and this was like a honeymoon cruise with his new one. Scary! On another occasion a passenger had died in his cabin. We thought it would be safe to move him to the hospital during the second sitting show. The other passengers would be in the show and in the lounges. One passenger was at the reception desk when we carried the deceased through the foyer, all wrapped up and neatly tied, and the passenger asked the receptionist, *"What's that?"* To which she replied, *"I think it's a surprise for the midnight buffet!*

There is always a lot of unprofessional jealousy when dealing with insecure people in charge. I was called to the radio room and told that I had to take over planning the next voyage until the new cruise director could find his feet. This had to be done secretly - Johnnie Smythe was going to be disembarked when we returned to Sydney. We were on our way to Hobart, Tasmania and I told Johnnie what was being planned. We went ashore and found a lovely little pub and the Tasmanian Devil took over. The pub was on the way back to the ship and bit by bit we captured each entertainer and staff member returning to the ship. This little pub had a drink called gellygnite. It was jelly in a little tub mixed with tequila and it filled up your cheeks and knocked your socks off. Ken Tweddle had just popped one in his mouth, I smacked his cheeks and this green mass jettisoned out of his mouth all over the table. Johnnie thanked me for my honesty and friendship and for a lovely send-off and together we all staggered back to the ship.

The new guy, Tim Skinner, took a bit of getting used to. But we just did what we did best and life went on. The next voyage was to New Zealand and we had never been there before. A world in one country is how I would describe it. Tom had injured himself and I gave him my cabin, which was more accessible than his was, and Tim Skinner moved me into a broom cupboard somewhere on Foyer Deck.

We arrived at the mouth of Milford Sound to drama on the 31st January 1991. One of the cabin stewards had a burst ulcer and a helicopter came to evacuate him. Then another helicopter came and dropped off a pilot that was going to take the ship into this wonderful Fjord for the day. I had never seen a helicopter on board and they pinned the Italian flag above the pool deck to use as a helipad. I helped Tom out of my cabin and got him comfortable on the promenade deck to watch the spectacular scenery unfold.

It was very windy with bits of rain which just added to the *4000* plus waterfalls that surrounded us. I stood with Captain de Rosa up on the bridge and he gestured to the pilot. His hands were shaking with the espresso coffee he was trying to swallow and pass orders at the same time. He made a mistake and would've taken the arse off the Achille, had de Rosa not intervened and changed the order from port to starboard. He ordered the pilot flag to be taken down and said to the pilot...*I will take my ship, you just show me the way!* The Sound/Fjord is too deep to anchor and we turned around at the end and came out safely under the command of the Captain. The helicopter came back and the pilot was taken off. Down in the South

Island, the nearest other land mass is the South Pole. The water is freezing and we were just about to get up steam when one long blast of the ships siren went off. *Man overboard!*

We raced to the back deck and there was a cabin steward floundering in our wake. The rescue party lowered a life boat (we hadn't seen that before either) and with our help from the elevated back deck, we gave them directions over the swells to the freezing cabin steward. They plucked out the little ice-block and brought him back to the ship. He was demanding to go on leave and the purser wouldn't comply, but he listened now! The helicopter came back again in case the fellow had hypothermia, but they stuck him in the hospital and probably banged him into a microwave to thaw out. The helicopter flew away and the Italian flag was returned to the bridge. The tannoy suddenly sounded again… *Robin call 320!* I phoned the bridge thinking, *what the hell have I forgotten to do?* The Staff Captain, Sylvestro Gentilliomo said the Captain wanted to see me in the Sorrento Bar. I raced to the bar and he was standing in front of Giorgio. They were both armed with a drink and there was a beer poured for me. The Captain handed it to me and we raised our glasses and Captain de Rosa said, "I *will never forget your birthday!"*

Allen Foggit phoned me in Sydney and asked me to try and get some staff together for a ship that he was planning to bring over to South Africa after the Achille had finished its season. At sea you miss out on land happenings. (No news, TV and CNN on our ship) No ship was travelling in the Med because of the Gulf War, and he got one to replace the Achille which next year was to cruise the Australian Waters. He got the thumbs up from our regular team but we first needed a break. When the ship went to Noumea, Terry

Fortune, Neil and I went on a 10 day break. Terry organised a few shows on land and I did a few with him. I phoned my friends Steve and Cara and they introduced me to a beautiful pharmacist friend of theirs called Julie. I took her for dinner at the Bourbon and Beef Steak restaurant and spent the next 10 days with her.

Neil was going to do our sound and lights and we met up to do the show. He and Terry couldn't believe I had this beautiful girl on my arm. We did the show and it was packed with Aussies and ex South Africans whom we made laugh and cry and made them all feel terribly home-sick. Terry did what is known as a false tab - you pretend to be finished and then come on to do a finale (if they wanted one). Yes, they did and I brought Terry out in a state of undress. I announced that tomorrow we will rejoin the Achille Lauro (Afrikaans... *beter bekend as die Archie le Roux)* to sail home to South Africa. Terry sang a well known Australian song, also composed by Peter Allen - *I'll rather leave while I'm in love.* Standing ovation and not a dry eye in the house!

Captain de Rosa was going on leave and I would never sail with this Gallant man again. Captain Giuseppe Orsi returned and the poor carpenter, Patilorro set about making new *no entry, do not pass go* and *f*** off* signs. Jealousy raised its ugly head. The Chief Purser, Aldo Accardi, used to show me torn up comment cards from passengers that Tim Skinner had discarded, but the cabin steward used to go through the bin. Some of them said wonderful things but one he had pieced together said - *What this ship needs is a few Robin Boltman's!*

We were having farewell drinks in Phillips Foot bar in Circular Quay when Neil came in and said we had been fired! I turned to Julie and said, *"Can I stay with you?"* We had an hour to pack and get off the ship! We just took the bare necessities and the crew were in an uproar. Tim Skinner was nowhere to be seen and our air tickets had been booked about two weeks earlier. The wonderful Mario, head of the Israeli Security on board, said we will see you soon. Sean, Neil and I were on our way home and Geraldine and Jane were in tears. Apparently the owner of MSC, Senor Aponte would be on board for the Sydney to Fremantle trip. Tim Skinner didn't want us around to make any comparisons. What a dip-stick!

The company put us up in a lovely hotel in Sydney for our last night and we ran up one hell of a tab in the restaurant and bar and Julie stayed the night with me for the last time.The next day was off to the airport and I had a voucher to fill in for the cab driver. It had a limit of $200 and the bill only came to $45. I asked how the tip works and he said just add it on because the meter will deduct everything. I said, *"Keep the change."* He thought it was Christmas. *F***-em!* On the plane Sean said, *"I've never been fired with my boss and assistant boss before!"* We stayed in Hong Kong and on to Taiwan and Mauritius on our way home. (I've just finished paying off the beer we had at the Peninsula Hotel in Hong Kong 🤣!)

I was in time to see John Wainwright one last time before he died. My first day back I went around to the house and Johns eyes lit up, but he couldn't move. Mandy phoned in the wee hours of the morning saying John had passed away. I was a pall bearer at one of the largest funerals I had ever attended. There were loads of entertainers there, and how Peter Taylor ever got through the song with a lump in his throat, I will never know.

Tim Skinner was disembarked in Bali and I had to fly to the Seychelles to re-join the Achille. Five days in Paradise, while I waited to rejoin the ship. I flew business class because the plane was full. Up in the bubble of the Jumbo Luxavia Jet, the passenger next to me asked as to what I did that I was so important, because these seats are reserved for VIP's. I said I wasn't a VIP, I was the Cruise Director for the Achille Lauro and he laughed and said he would come down and have a drink on board; he was the harbour master! All the air-hostesses were Marisa's Luxavia friends and I was looked after second to none. I had a few lovely days on the Island and sat at the Coral Strand Hotel pub watching for the ship to break the horizon. It was very late and Tom explained later that the Helmsman in the night had made a mistake and did a contra bearing reading and turned the ship around! One day Tom also turned the ship around...

From Durban to Cape Town, a passenger asked Tom, while he was playing the piano one evening, if that land on our starboard side was Cape Agullas. Tom wasn't sure so he phoned 320 and there was no reply from the bridge. He then dialled 244 to the reception desk, and got hold of a Swiss/Italian called Kersey. (He had lovely nails!) Tom queried as to the correct number for the bridge and Kersey confirmed that it was 320. Tom said that's not good enough that the bridge doesn't answer their phone..."*There could have been a man overboard!*" he said. Tom then phoned the bridge again, got through, and confirmed the land was definitely Cape Agullas on our starboard side. Tom returned to his piano and told the passenger, pointing over to the starboard side, that it was Cape Agullas and it was gone! Kersey, being Swiss & Italian, didn't understand Tom too well. He phoned the Captain in his cabin and said, "*Tom said there is a man overboard!*"

One long blast on the ships siren and they executed the Williamson's turn to get us back to the position for the man overboard. Officers came running into the Capri bar and asked Tom, back at his piano, for the man overboard. Tom said,*" I'm not sure of that one, but hum a few bars and I'll try and play it!"* Tom was summoned to the Captains office and immediately congratulated the Captain on a brilliant manoeuvre... *"We didn't even lose one glass in that turn."* The officers of the watch got into shit!

It's difficult to keep a secret on board. Ken Tweddle had got married in a quiet ceremony before we had left Australia, but his trombonist had to be the witness. Tom Trombone told me about it, and the night before we arrived in Southampton, we were doing the Farewell show. Ken called for *I still call Australia home* and the band played the *Wedding March!* His face was a picture, as I walked out with his new wife and all the entertainers, with the largest wedding cake the galley could bake. It was a wonderful way to finish the season with the Achille Lauro and our Australian friends.

Julian, Neil and I had a few days in London before we flew to Paris, Dubai and on to the Seychelles to join the Oceanos. Jannie, Lorraine and Geraldine were there already with the Carlo Spetto Dancers... Kerry Menzies & Nikki Bullock and her sisters, and a few other entertainers I'd worked with before in various shows around the country with Sun International. All the rehearsals were done in the hotel, while Neil, Julian and I tried to play golf in 40 degrees, waiting for the arrival of the Oceanos. Jannie was clever... he took all the unopened airline lunches and dinners that hadn't been used and took it with. He'd been in situations like this before and he was right again. They came in very handy!

MV OCEANOS

A lovely little ship that was Greek owned and was a lot smaller than the Achille; only 16000 tons. We sailed off to Durban and Tom, Terry, the famous Tony Schilder (you met at the holiday Inn, remember?) and Alvon Collison were also on board. I used to visit Alvon at the Cape Sun regularly when he performed there with Taliep Pietersen. Tom's cabin, up forward, was so small that he was the only one without a telephone. There was no room! He had to open his cabin door just to be able to open the daily program. The Greek captain was a real smoothie and apparently used to work on the *Love Boat.* (Not during the filming!) That was the theme the band played for Captain Yannis Avranas at every Captains Cocktail party. He had his wife, Ingrid and their daughter, canaries and his dog, Cooper on board.

One day while I was doing the Lat. & Long during the noon report from the bridge, Cooper bit me. There was a commotion over the tannoy system and I eventually came down with my hand bandaged up. The passengers wouldn't believe me! It does sound strange to be bitten by a dog three hundred miles out at sea! Kevin Savage came on board along with Martin and Pam Locke. Kevin and I had worked together during a few *Late Night Live* shows. They were going to do a live link with Radio 5 from the ship. The technical staff won an award for the broadcast. Remember this is before laptops and cell phones. During the broadcast Kevin interviewed me and asked about the captain's cocktail party on deck, *"What's in this cocktail?'* I replied, *"It's a mixture of witblitz, after-shave and windowleen. It makes you drunk, but you smell great and your eyes will be clear in the morning!"*

I got off in Durban and flew to Johannesburg to buy a new sound system for the pool deck. The ship went to Mauritius and Reunion and I rejoined the ship in Durban. While we were tied up, the tide went down and the huge tyres attached to the quayside lifted the door off and down it went. Divers were sent down to retrieve it before the passengers came aboard. Imagine coming on board and we were without a front door? We sailed to Cape Town and I visited my family. My dad was very ill and when I got back to the ship, Tom and Terry asked why I was looking so glum. I said, *"Have you ever said goodbye to someone knowing that you will never see them again?"*

My dad died and TFC were telephoned and they explained to my mom that there is nothing we can really do at sea - which is true - it would just spoil the voyage for the passengers and me. As an entertainer they would miss out on shows, etc. I knew when I got the message in Durban to phone home. Lynne and Geraldine were so helpful and Graham came to fetch me, and I left the ship again. I flew down to Cape Town and during the funeral I felt a firm hand on my shoulder outside the church. It was Mario, boss of the Israeli Security from the Achille. He had read about it in the newspapers... friends!

This next entry into my story is courtesy of my friend Andrew Pike, the author of "Against all Odds" - the epic story of the Oceanos Rescue.

During my absence the ship had sailed to the Island of Reunion and on board was a fellow called David Gordon. A leading senior advocate at the Durban Bar, with a speciality in maritime law, he was a veteran cruise ship passenger and knew that the way this vessel was rolling on relatively calm seas was unusual for a passenger ship. Once on board, he re-engaged his interest in cruise ships, wandering around the ship filming everything he could see with his video camera, including lifeboats and safety equipment and finding very little to his satisfaction. As the ship sailed out of Reunion David and a colleague were standing on the port side of the vessel, with David filming the ship's departure. Suddenly his colleague exclaimed: "We've just touched the bottom!" David pointed his camera downwards, towards the water. Mud and sand could clearly be seen, churning up close to the vessel's port side hull.

That evening David and his wife Anne dined at the Captain's Table, something of a privilege on a passenger liner. During the dinner David asked captain Avranas, "Did we hit the bottom when we sailed out of Reunion harbour?" The captain replied in his thick

Greek accent, "It was nothing. We touch an underwater buoy." "Why would there be a buoy underwater? Is it for the Reunion submarine fleet?" David asked facetiously. "Surely the vessel touched the ocean floor?" "It's nothing," said Avranas. "I know what I'm talking about. You doubt me?" Avranas's arrogance cut short the conversation and the rest of the dinner continued uneasily. (Quoted with Andrew Pike's consent from "Against all Odds") Mandy Wainwright, John's widowed wife was also on board and she also filmed the drama.

I flew back to Durban and the ship to hear that Allan Foggitt had sold TFC to another company. We got under way to sail to Cape Town. July and August is not a good time to experience the Cape Rollers. There were Skips little tin-foil bags all over the ship and the staff would warn each other of dodgy spots on the carpet by saying, *"Staff call... Pizza alert near the casino,"* over the Tannoy system. There is nothing worse; I'm led to believe, than being sea sick. I've sailed through cyclones and never felt it. I used to tell the passengers that it was all in the mind, sometimes all in the stomach, but nine times out of ten it was all on the floor!

My mate Sidney and I went halves on purchasing a shotgun, ammunition, clay-pigeons and a launching trap to do skeet shooting on the deck. I had cleared it with the Captain and I looked up the barrel, put it in my wardrobe and locked the ammo up in the captains safe. Sid and I were going into business! Our business was very short lived - we were going to go into "Liquidation" very shortly!

When we sailed out of Cape Town I said to my mom that I felt sorry for these passengers. The weather was dreadful and we battled for over an hour trying to get the pilot off outside the Moullie Point light house area. The ship actually turned to face the lighthouse to create a lee for the Pilot to disembark onto the pilot boat. The port authorities closed the harbour after we sailed out. Our passengers went to the Holiday Inn and Fish River Sun when we arrived in East London. Winston Sahd had chartered the ship for two days and had a company do on the first day. The second day was for his daughters wedding. Winston wanted entertainment every hour on the hour after the ceremony. I drew the short straw and I went on stage about 5-00am that morning. All the staff worked hard during those two days. We sailed out for the wedding ceremony, but came back to the harbour because the weather was so bad.

I saw Winston again years later, at the Westlake Golf Club during a show, and he said that the marriage didn't last... I replied, "Neither did the ship!"

Our old passengers from Fish River Sun came back and we also had a lot of SAA staff on board. One air hostess left her friends on board saying that she refuses to *even fly* in weather like this! We did the sail away party in the main lounge because the pool deck was impossible. The awning over the pool bar had already blown away, along with trees and bus shelters in East London. We did our best falling all over the stage, but when Terry and Julian landed on top of me, we decided to call it quits! We had one show to do that night; Terry Lester in cabaret where I was going to be the MC and then we could all have an early night. We hadn't slept for two days because of Winston's party. We weren't going to sleep for another two days... it was the 3rd August 1991.

I'VE GOT THAT SINKING FEELING

While the passengers were having supper I was talking to Nikki and the barman, Paniyotis, in the Four Seasons lounge. A crew member came in wide eyed and shouted something in Greek to Paniyotis. He rushed out and asked me to watch the bar for him. I thought there was a fight downstairs amongst the crew and Paniyotis was a big, strong lad. He came back later covered up to his elbows in shit. They were battling to close a valve near the generator room; he couldn't close it either. I went to my cabin to prepare for the show and all the lights went out and the engines had stopped. There was just an eerie glow from the emergency lights in the passages. After bouncing about in the cabin while trying to get ready for the show, I raced up to the show lounge in my suit and tie because I knew all the passengers would be there for the show. Lorraine came up and gave me the megaphone and said to keep them busy and to keep them calm. I got my guitar from backstage and the first song I played was Jeremy Taylor's, *Ag Pleez Daddy* and they all joined in. Then Moss, Julian and Tom got onto the stage and we played some more... *American Pie* is a great sing-a-long, except when you're sinking! When we got to the line, "*This will be the day that I die*" we decided to change the song! Tom left the stage when his piano and the drums left the stage, yet we continued singing with my banjo which Moss had retrieved from the downstairs lounge, so that he could play my acoustic guitar.

No alarms, no announcements, no seven short blasts and one long on the alarm bells around the ship. When the shit hits the fan the rule book generally goes out of the pothole! Kevin Ellis, the dancer, was the first one I saw with a life jacket. I told him to take it off but he said all the crew downstairs all had theirs on! Then the idiotic Staff Captain, wearing his life jacket, came storming in and screamed to everybody to lie down; just like he was about to hold up a bank! I told him to get out in stronger tones than that - I had just managed to get everybody calm and he came in causing havoc! He had a twisted arm from a previous break, and now I knew how he had got it! Obviously from some other sinking drama! I never saw him again. Some of our staff and crew went down to get life jackets from passenger cabins to bring up to the lounge.

Downstairs had been declared out of bounds to the passengers and Tom had stationed himself at the top of the stairs to prevent the passengers from going down below decks. If they had seen the water downstairs they would have panicked. Panic is contagious, and so is laughter. I told the staff to walk around and smile like dolphins and John Denver.

The passengers were amazing and so were the TFC staff and entertainers. Unbeknown to us the life boats were already being lowered. A passenger came and asked as to why the lifeboats were coming down? I said it was probably a precautionary measure. Moss had been filming downstairs and came to tell me that below decks, the place was flooded. *We're going down!*

The life jackets arrived and everybody put them on. These passengers had already done two life boat drills, but we had a handful of new *pax* on board. I asked that if anyone was unsure as to how to wear a life jacket, just raise a hand, and we would show them how. Most of our new comers were SAA (South African Airway's) staff - *they could show us!* I went out to the decks and couldn't believe my eyes; crew and suitcases were already in the life boats. There is nothing like being forewarned! We got as many passengers into the lifeboats (crew permitting) and the SAA staff we tried to distribute evenly amongst the passengers - they've been trained for emergencies. I missed Neil then, he was on holiday and an ex policeman! The Captain put his wife and daughter into a life boat with Cooper (his dog) and then tried to get in himself. Lorraine and Geraldine prevented that. After that he was useless to us... *women and children first!* SOLAS (Safety of lives at sea) have since looked at and changed this. Dads must go with! Imagine only women and children on a life boat with no dad to help them? What goes through a child's mind when the family is separated and they are crammed onto a little boat in the dark? Anyway dads will be useless on board because their mind would be out on that little boat! Just like our Captain's!

Lynne was in one of the lifeboats and she had a radio with her. There were between eight and ten meter swells with a howling wind. To launch a boat in these conditions is nearly impossible. The ship is rolling, without steerage, and is at the mercy of the sea. Once the life boat hits the water, it has to be released as quickly as possible. The swell goes up and then it goes down. The steel cables are taught and then they are slack, and hooks are flying about the passenger's heads once they have been released, as Lorraine's poor Granny found out. She took a gash to her head but never complained. She was as tough as Lorraine. There was a ship nearby, the Nedlloyd Mauritius, and Lynne was trying to head for it in the dark. She wanted directions over the radio but we weren't sure where she was. *"Let go a flare"* I told her, and we waited. *"Well, did you let one off?"* I asked. "*Yes, but it shot into the f***ing water!"* she replied. She was very brave and only she can tell you, along with her fellow passengers, what it was like to spend the night, in wild seas in a life boat.

We had our own problems on board! The life boats had all gone and we still had 270-odd people on board. They were calm and the battery in my megaphone had long since died. Just as well, because some people had fallen asleep! That's how calm they were! Alvon accompanied me for a while telling jokes to satellite groups that were still awake. One fellow told us a joke and Alvon said, "*What a kak joke! For that you're that we'll keep you on till last!*" I also noticed that nobody was sea sick anymore... they were too busy shitting themselves!

A passenger came up to me to get permission to go down to his cabin to retrieve his Durban to Johannesburg air tickets. He had a very important meeting the next day and I said, "We're not going to Durban, you're not going to the meeting, we're going down!" I went down one deck to the Four Seasons Lounge to bring up some soft drinks for the passengers. The bar upstairs was depleted already. It was happy hour, there were no bar staff, *or any staff*, for that matter! I offered Tom a beer on my way back at the top of the stairs, where he was still patiently stationed. He pointed to the pot plants. He had a six pack hidden, already on the go. He had also rescued our Astor beer mugs and had them safely tied to his life jacket. (I met with his sisters a few years later. They told me that his families in England were watching the news at the time. When they saw Tom getting out of the helicopter raising our beer tankards, they just burst into tears.)

The ship was listing badly to the starboard side and the piano had long since left the band stand. (of its own *A Chord!)* I discussed it with Tom and we decided to move all the passengers out to the pool deck. I didn't know how long it would take before we sank or rolled over. We would have a better chance of surviving outside than in the lounge. I passed Pat and Lionel Roche on the way out and he said he can't go into the water because of his pacemaker! I said we won't go in the water - this is the Wild Coast - there are sharks in here that bite your bum! He just smiled and I looked at my watch; it was about 3-00am. They were due to sail with us tonight from Durban on the next voyage, and I had arranged a surprise cocktail party for them. I shook his hand and gave Pat a kiss and said, *"Happy wedding anniversary!"*

Sitting in the lounge we had all become accustomed the listing of the ship, but when we got out to the deck, we could see how bad it was in relation to the horizon. I told Tom to stop walking about because of the angle of the deck. He looked at me and he said, *"Are you mad? I've spent most of my life at this angle!"*

Lorraine came to me and said the staff had found a life raft and a rubber duck in the bow to use as a last resort. We went down and we decided to tie it up because of the howling gale. I said to only launch it when it got light enough to see, but somebody should watch over it. Leonie promptly volunteered. I gave her my pocket knife to cut the ropes when needed and she jumped in and went to sleep! Lorraine said she was frightened but she sure as hell didn't look it. She was amazing! I then decided to check with the bridge and went above decks. The bridge was deserted! Just this little voice saying, *Oceanos come in... Oceanos come in..."* I answered the radio. It was the container vessel Nedlloyd Mauritius. They were busy picking trying to pick up some of our crew and passengers out of the life boats. One of the little babies went up in a fire bucket! I met him and the family years later in Kommetjie and thank goodness he remembers nothing about it. He wouldn't fit into a *barrel* now!

Then a new voice came on the radio… *Mobile 7 Rescue* was how they identified themselves. Silvermine in Cape Town had picked up our Mayday (and a few other "ears"… even the famous Robberg Radio in Norway) and the rescue wheels had been set in motion. They had set up a base at The Haven on the Transkei Coast near Hole in the Wall and Coffee Bay. The navigational charts had all slid off the navigators table and they wanted our position. I told them I can see two lighthouses from the bridge. Tell them to switch one off and I can give a contra bearing from the compass (The GPS was off as there was no power). On the other hand, I told them, there are a few ships around us, we are the white one with the blue funnel, and the only one that's nearly upside down! How F***ing difficult is that to spot?

We were on the Mayday Channel 16 which should always be kept on and free for emergencies. Once you've made contact the procedure is to change channels. I said that this radio is connected to a car battery… I'm not going to change anything. They wanted to know how many pax I had on board. I went back to overlook the pool deck. I estimated about 270 passengers were still on board. I counted ten and then went ten, twenty, thirty etc. They replied that help was on the way in the form of 11 Puma Helicopters and one Allouette. They gave me the direction via the compass from where they will be coming so that I could tell the passengers and cheer them up. I returned to the pool deck and they were already being cheered up. Dolphins were playing in front of the assembled crowd, and colours were just starting to take over from the darkness and overhanging gloom. With the other ships around us, why did the dolphins play here? Why not on the port side or at the front end of the ship? No, just in front of the passengers, who were sitting in long lines, with their legs wrapped around each other to prevent them from sliding down the steep slope of the deck.

Ronan, the shopkeeper, smashed his window (he didn't want to go below to fetch the keys), and handed out windbreakers and any form of clothing his shop had to offer. Aren't people fascinating... *"Can I have the blue one? Have you got this in a medium?" (F*** off comes to mind!)*

Anyway, I pointed to the coastline and the word was passed on... Then, over the sea at first light, two helicopters were skimming over the waves in formation towards us. The other helicopters that had flown during the night from Cape Town and Pretoria were being re-fuelled. The cheer that went up could have been heard at Water World at the Wild Coast Sun. I raced back to the bridge. We had a helipad in the front and we could do the airlifting from two places; forward helipad and the pool deck aft, where the majority of the remaining passengers were assembled. Two navy divers were lowered to the ship, Paul Whiley and Gary Schouler.

Geraldine,Lorraine, Moss and Julian moved the more agile to the front and before the airlifting could begin, I had to cut the fairy lights that hang over the ship. I clambered up the funnel but they were attached by a heavy steel cable. On channel 16, I asked for bolt-cutters and seconds later a helicopter hovered over the port wing of the bridge and lowered them down to me. These guys thought of everything! Then the airlifting began... two by two... forward and aft. The helicopters formed a queue... two over the ship and the rest behind. When two passengers were in the harness, the helicopter would fly off with them, winch them up and get back in the queue. When they were full they would drop them off at The Haven and come back and start all over again. I remember Alvon saying, as the dancing girls were being airlifted, *"I have to go now too, I think my waters broken!"* Hell, how he made me laugh!

Julian and Gary (diver) had launched the rubber duck off the front and were going to pick up people who were jumping off the bow, which by now was very low in the water. They would take their survivors to the Nedlloyd Mauritius lifeboat - they had launched one of their own lifeboats in this weather, and returned for more. I was on the bridge trying my best to give Mobile 7 an accurate count as to the survivors that were leaving the ship to be rescued by the rubber duck and the Nedlloyd's lifeboat. Then a new voice came on the radio... it was our Captain! He was already at Mobile 7! *"What degree are WE listing at now, Robin?"* I thought about it for a moment, "W*hat do you mean, we?"* I checked on the inclinometer and it only went as far as 30 degrees. As accurately as I could with a pen, I filled in the little spaces trying to follow the same arc as the pendulum. "It's about 38 degrees!" There was silence from the other side. There was silence from outside too! The helicopters had gone back to refuel. "*We will be back in two minutes"* they said. (Two minutes is not a long time, depending on which side of the toilet door you're standing)

I let the Captains canaries out of their cage and showed one of them, sitting on my finger to Moss down on the Helipad. Moss looked knackered, but he managed a smile and a thumbs-up. He had been busy putting people into the harnesses for the air-lifting. Mobile 7 came back on the air... refuelling was nearly finished and how many people did I have left. I counted and said 14. *"Is that with you? No, make that 15!"* When you guys come back, I told them, don't go to the front, it's nearly gone. I'll take everybody to the pool deck at the rear.

Do a last check around the ship and in all the passenger lounges, they said, and come back to the radio. I knew the lounges were all clear because during the night, I took Hilton Schilder with me to do a look round, and as we passed the bar, he took a bottle of Scotch off the shelf. "*I've never been able to buy you a drink, so Cheers!"* Nothing we drank that day or night had any effect! Terry Lester told me later that if he had known we were going to sink, he would have run up a bigger bar tab!

Back on the bridge, the ships around us asked permission to leave the area. I thought to myself, "Who am I to tell these people what to do?" They replied, "You're on the bridge, it's your call!" I thanked them and told Mobile 7 what I was wearing and not to leave me behind. I still had to make my way back to the pool deck. Paul Whiley was there, all wet! Someone had fallen out of the harness and he had dived over the side, got hold of the passenger, took him to a life boat and swam back to the ship. He stayed behind to do a final check and I went up alone in the harness. Over the Dolly Parton Tits of the life jacket, I looked down at the ship. The props now out of the water and the bow was under water. What an awesome sight! On board we had become accustomed to the listing of the ship, but from an offshore point of view it was something else.

When we touched down at The Haven, two Army officers from Mobile 7 slid the helicopter door open. *"Who's Robin?"* I said, *"I didn't do it!"* (Old childhood/Army training reply) *"Come with us"*. I had to do a de-briefing while everything was still fresh in my mind. I was now going to meet the voices I had been talking to for the last few hours. When I walked into the Radio Room, the Officer in charge looked up, and with a grin said to his mates, *"Fuck it - he's still wearing a tie!"* My first words were, "Has anybody got a light?" My Zippo had died and I was a smoker - I even smoked in between smokes! I'm trying to cut down to one at a time. For the last eight hours I had had not one cigarette!

I sat and answered questions for what seemed like ages, while they wrote and recorded everything. All I wanted to do was to see the mates & Nikki, etc. Meanwhile, Tom and Terry were in a state of panic. I was nowhere to be found. All 270 odd survivors at the Haven had filled in their names and next of kin and my name was not yet on the list. (The Captain was no. 7) They got hold of an officer from Mobile 7 and he informed them as to my whereabouts. When I eventually joined the gang it was a great reunion. I saw the Captain, standing very much alone. He had already been recorded on camera declaring that, "When I order abandon ship, it doesn't matter what time I leave. Abandon is for everybody. If some people like to stay, they can stay."

This interview remains on social media to this day… and ignores the fact that neither he nor any other member of crew ever gave the order to abandon ship. When I approached him he looked down. Remember he had spoken to me on the bridge a few hours ago. I took off his windbreaker that I had found on his chair on the bridge. I handed it to him and said, "I'm sorry about your ship. I'm sure your wife, your child and Cooper (the dog) are fine because the rescue ships told me on the radio that they have taken all survivors out of the life boats. I've also set your canaries free." When he looked up, the tears just rolled down his cheeks as he shook my hand. He didn't have anything to say and I never saw him again.

The Haven is a wonderful small resort on the Wild Coast, near The Hole in the Wall and Coffee Bay. Talk about the 5 loaves and 2 fishes, (not during the sardine run) they served everybody soup, tea, coffee, etc. with the limited crockery they had. Hats off to The Haven and staff, they were just great.

The rescue officials were trying to get everybody listed, as were the ships at sea with our other passengers and crew. (Sorry, wrong order, crew and a few passengers) Dads were worried about wives and kids, business men were worried about their secretaries and what their wives were going to say... I bet some of those guys wished they could go back to the ship! But unfortunately, The Oceanos was no more; it sank just over an hour after I had gotten off.

Military buses arrived and we thanked The Haven crowd for their hospitality and catering and we all got aboard. I don't recall much of the journey back to East London, because I slept most of the way, as did most of the team. It was the first bit of sleep we had had in nearly four days.

We arrived at the Holiday Inn in the evening to be met by police road blocks and pandemonium. The public were being kept out and then the press interviews started, along with every available incoming telephone line. We did interview after interview well into the night. One of the hairdressers called me to the phone in her room, it was an English Newspaper, they wanted to talk to the magician on the bridge. When they asked what part of England I was from I told them I was from Cape Town. They were only interested in British role players. Anyway, much to their annoyance, I hung up. Let them battle to get through again.

To this day I'm very skeptical about what I read in English papers and see on BBC/Sky - it's very one sided. As it turns out, according to BBC footage, it was Julian Russell on the bridge *and* in the rubber duck for the last few hours. (That's a good trick! *If a man could be two places at one time, I'd be with you…*) That, I'm sure was his Dad's mistake as he told the press that his son was the magician on board, not knowing there were two magicians on the ship. There were numerous documentaries made, but M-net's, Canada's and America's Savage Sea documentary, and National Geographic's were the only accurate one's I've seen.

M-Net's documentary had the video footage that Moss had recorded of me on the bridge when I quoted the famous line from the Oceanos drama, "*Gooie More Suid Afrika, die luitste nuus is, ons is almal in die Kak!*" (Afrikaans for, "Good morning South Africa, the latest news is that we are all in the shit!") from a current morning live television show at the time called "Good Morning, South Africa" Bill Faure, the pioneer and CEO of M-Net (older brother of Duncan Faure of *Rabbit* and later *Bay City Rollers* fame) said that they would have to bleep out the word "Kak" for the documentary.

I told Bill that the show wouldn't be authentic, at a time like that people don't say things like "Jumping Jimminy, Goodness gracious me" etc. The other guy's all agreed and Bill relented. They used that snippet to advertise the upcoming show to be aired soon... and instead of hearing it just once, viewers heard it "Forty" times... I was the first person to say, "Kak" on National Television!

Julian and Moss were interviewed on an English One Sided Documentary and I'm sure they were told not to mention the South Africans. They interviewed Julian as the ship's magician and used footage that Moss filmed of me on the bridge and pretended that it was Julian.

(We both had blonde hair wearing a life jacket; they just took my Yorkshire? Afrikaans, South African accent out of the recording... and my name!)

I was nominated ABC's Personality of the Week Special and later I was flown to the United States to do an interview on the Savage Sea Documentary. I also received a wonderful letter from Brig. De Munnick from the SA Navy at Silvermine and was given an honorary membership of 22 Squadron in Cape Town from Col. Mac McCarthy.

Terry Lester, Moss, Julian and I stayed in East London to finish off all the interviews while the others all flew back to their homes, courtesy of SAA. Then the big news broke... everyone was accounted for! Not a single life was lost! My mom recalls shopping in *Pick'n Pay* when the news on the radio (KFM) was announced. She and other shoppers burst into applause. Needless to say my Mom filled the trolley with tears. Terry, Moss, Julian and I had a million drinks and staggered off to bed; singing at the tops of our voices, arm in arm, and while outside, the sea was as flat as a pancake!

Later during investigations by the various departments, David Fiddler (Principal Officer, Department of Transport, Maritime Division, Durban) Interviewed Chief Engineer Panayiotis Fines (*Takis*). He had actually brought copies of the deck plans that he had salvaged from his office before taking over his assigned life boat. "Takis" had proved to be more co-operative than Avranas. David Fiddler had pondered over the plans. He knew that, as a car carrier, the bottom deck would have been just above the sewage tank. Takis had gone into the generator room after hearing the "Bang". The water had got into the room. He had been trying to push the sea chest back onto the hull. He had scars and wounds on his arms. The ship was a converted car ferry and the deck above used to house cars, and now it was converted into cabins. Those cabins always had passengers complaining of the sewage smell down there. The venting pipe was often removed for cleaning purposes (hence the big barman Paniyotis's sleeves covered in shit). When Takis had left he closed the water tight doors and left. David Fiddler raced back to the hotel where Chief Engineer "Takis" was and David asked him, "Chief, tell me truthfully. Was the venting pipe in place when the sea chest came off?"

He hung his head and replied: "No, it was not. We remove a part of it." "Why not," asked David? "That pipe, it was often blocked. The passengers complain of a shit smell. I always have to remove pipe to clean it. When the water come, there was no time to replace."

"So you knew," continued David, "that once the water in the generator room would pass through to the engine room when it reached the level of the venting pipe hole. And you knew that the water would flood forwards through the ship and it would sink after

THE SOCIETY OF AMERICAN MAGICIANS

MAGIC · UNITY · MIGHT

THE HEROISM AND PATRIOT AWARD

...sm and Patriot Award

Presented to

Robin Boltman

Robin Boltman is a magician and a hero August 3rd and 4th 1991 Off the Wild Coast of South Africa onboard the sinkin cruise liner Oceanos, Robin Boltman the ship's magician entertained and calme passengers and maintained radio contact with rescuers while awaiting rescue. When the helicopters arrived at first light and Robin was asked to remove the overhead fairy lights that was preventing the airlifting operation, at great personal risk Robin climbed up the funnel and cut the cable. During the airliftin Robin directly assisted the rescue operation saving 571 lives. Robin was the ne to last person to ever leave the Oceanos, Forty five minutes later the luxury line disappeared under the waves of the Indian Ocean. Nov 30, 1994, onboard the Crui Ship Achille Lauro burning 125 miles off the Somalia Coast; Robin again played a major role during this rescue. We honor Robin Boltman with this award.

Presented July 16 2011

Mark Weidhaas, S.A.M. Presid...

you closed the water tight doors?" Fiddler pressed him. "Yes, there was nothing we could do. There was no time." *(This technical info courtesy of Andrew Pike's Against all Odds)*

Back to safety and waking up in the Holiday Inn in East London, the phone rang bright and early the next morning. Sue Kelly-Christie and then from Danny Fisher... his girls wouldn't go to school until they knew that their uncle Rob was safe.

We had a reunion party with TFC staff and another with Irene Smith (Alex Jay's mom; he's a popular DJ in SA). His mom and sisters were also survivors. The South African Magic Council gave me a huge party with members present from all over the country and made me an honorary member of all the Magic societies in the country... along with a beautiful gold medal inscribed... *"For bravery and outstanding service to Magic".* Members from all over, including some of my mentors, like Graham Kirk, Martino and Brian Marshall. There were a few tears there too! (Mostly from me) I was also awarded a Medal from the American's... *"Heroism and Patriot Award"* for entertainers. Recipients include Fire Fighters from the New York 9-11 disaster, Afghanistan and Middle East soldiers in action, etc. To date seven medals have been awarded and I am the only non-American to receive it. I was a very proud boy!

I went home to Cape Town to see the family, and at the airport, my family had gathered all the friends and relatives...it was like being at my own funeral, only happier, with one extra drunk!

The next day, my pal Sidney and I did a pub crawl. From Forries to Hout Bay, loads of stops along the way, and we never paid for a drink. There were so many well wishers, slaps on the back and finally I arrived in Table View at Derek Gordon's Pub, "Smilers". He was on stage and announced I was in the room." *Would our hero please stand up and take a bow?"* To a wonderful applause I stood up and promptly fell over, arse over kettle!

I did quite a bit of TV work after the ship and one memorable show produced by Anne Williams and hosted by Clive (*Ted Dixon* of *Villages fame)* Scott was a real hoot. It was a New Year Extravaganza and would run from an hour before and after midnight. It was recorded on a Monday in October and there was a huge cast of entertainers. The dressing room next door had Casper de Vries and Mark Banks. They had an ongoing game of moving each others names on the door… Mark on top, Casper below… then Casper would put his name right at the top of the door and Marks right at the bottom. Mark would return and tear Casper's name in half and replace his in the middle… Eventually both names hung semi burnt on the door.

During all the rehearsals the Cape Minstrel Follies had ongoing music down in their dressing room. Barry Hilton and I went in and I borrowed a banjo and we all sang, "*For you are enchantment.*" Not many people know this Cape District Six song! Later back in our dressing room, quite a few of the gang were all trying to get into New Year mode (in October) and the Minstrels were still going hell for leather down the corridor. Malcolm Terry came in and with his hand upturned on his forehead and said, "*Are those bloody Follies ever going to shut up?*" An instant reply from Mark Banks, as only he can do was, "*What's the matter Malcolm, are you on a Minstrel cycle?*" It was the best line and it wasn't in the show. Anne Williams won an award for the show.

THE TOUR LEADER

While we waited for the next ship to replace the Oceanos, I became a tour leader for John and Allan Foggitt. It was a fascinating trip and we had a lovely guide in her 60's, fit as a fiddle, called Madame Fatma Rifki. We cruised down the Nile, visited all the Temples, the Pyramids and the Sphinx, and I made a point of checking out the bridge... just in case! I also had Pat & Lionel (Oceanos) with me. They insisted I perform for these passengers and informed the Crew as to who I was. Decked out in my Egyptian Galibaeo I entertained them quite often. This would often spill over to the various hotels we stayed in.

We visited the War Graves and I joined a passenger, Ken Wade-Lehman from Durban, and his mate, to visit a grave of a friend of theirs. They were in North Africa during WW II and I watched from a short distance as these two sat down and chatted to their fallen Comrade after all these years. They spoke about when they had last had breakfast together and explained to their fallen comrade as to who was responsible for delivering his last message's home, who took his blankets and who 'borrowed' his boots. It was a very sombre coach that traveled back to Alexandria that evening. Then we flew off to Israel for the next leg of the tour.

We saw some lovely sights in Israel, Masada, the Dead Sea, The church in Bethlehem and the Wiling Wall... I was the only idiot with a harpoon! In Israel one evening, my sister Beverley, Avi and the family came to visit. After they had left, while waiting for the passengers all to go to bed, a beauty came over to me and invited me to join her and her mom in the lounge. In this Jerusalem hotel there was no pub... you drank at a table and a waiter served you.

She had a baby with her and shortly she opened her top and started breast feeding. "I hope you don't mind, but babies can't tell the time". I said I didn't mind and I would love to do that myself, but the baby probably wouldn't like my tits! They both laughed. I was immediately up for adoption! Mom took the baby to bed and we sat together having a few drinks. I was very curious about this girl because everybody seemed to know her. They would wave and smile and nod. At first I thought it was me as I had just been seen on nearly every TV news item around the world because of the Oceanos, and I had felt like Frank Sinatra... well for about three day's anyway! I asked her who she is and why does everybody know you. She said she was Miss Israel but lost the title because she was pregnant. She was still Miss Israel to me!

I did the Egypt and Israel trip twice and Egypt remains one of the most fascinating countries in the world. I performed on the Nile dressed in a Galibaeo and introduced many of the ship board activities to the crew. But instead of staying to wait for the third group, the phone call from Diane Head from Starlight Cruises said the new ship will be waiting in Durban upon my return. Come Home!

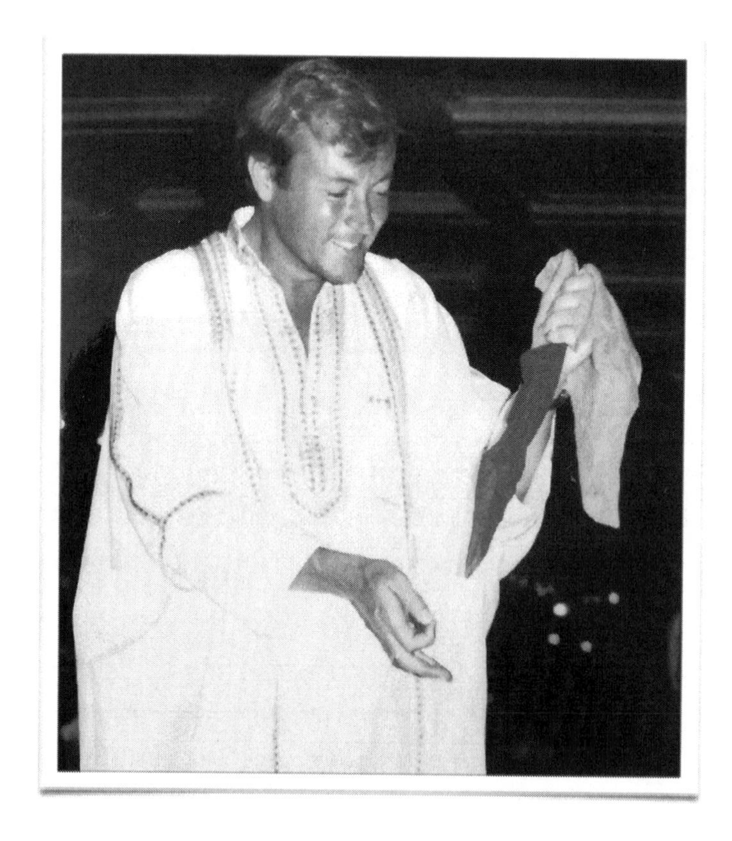

AEGEAN DOLPHIN

With the gang already on board, and a few new faces, we set off to complete the Oceanos' program. The wonderful cabaret singer, Ann Powers was on, along with Garth Victor and his band, a great bunch of muso's (Lionel Bastos is often on the radio). Sean Anderson informed me of a barman that worked the pool deck and Tom's piano bar. Apparently he didn't like the cruise staff and kept them waiting for ages. The cruise staff also had an allowance to reciprocate drinks bought while they socialised. Bearing in mind, this Greek crew had a hard act to follow, what with some of the "heroes" of the Oceanos being on board, and passengers' comments after life boat drill, *"Have you got your life boat ready? Packed, are you?"* It went on and on.

I headed up to the pool bar, introduced myself to the problem barman, Kiriyakos, and ordered a packet of Marlboro cigarettes with a $50 bill. He went off at me about his cash float and I told him that I didn't want any change. I never waited for anything again in either bar! When Kiriyakos spotted me crossing the pool deck, my Castle beer was poured and ready.

Jannie and Nadia were Cruise Directors and they did a wonderful job with the limited resources that the new Company had at their disposal... next to nothing! We had a wonderful band, Garth Victor, Lionel Bastos, Tam Minter and some others. Clem Leach was a new host, the son of the new owner, and he had the rhythm of a fridge magnet. He did his best in everything and staff and passengers couldn't wait for Clem to do his bit during the staff shows.

At the end of each voyage, it was tradition that the Cruise Hosts and Hostesses did The Cruise Show. It was a variety show combined with everything we knew from the Astor days to date. To see Clem dressed as a ballerina during *..."If I were not upon the sea..."* it was the most hysterical sight. Even the cast in the line-up used to fall about.

I got involved with a hostess called Lindsay McKenzie, a child hostess. Between her and Bronwyn Lazaar, the kids were in safe hands. After a late night in the disco, I popped in to see them. Both of them were fast asleep with the children face painting anything that didn't move, including Lindsay and Bronwyn.

I received a phone call from Jose Broude in Johannesburg saying that Sun City and Carlo Spetto were putting on the new Extravaganza and they wanted me in it! I would be the first SA born specialty act to be in this show. (My second appearance actually - I stood in for the American magician called Mike Michaels who had another contract to fulfil) We all celebrated around Tom's piano that night - with Kiriyakos! Some good was coming out of the Oceanos... if not my belongings.

Moss had written a song about the Oceanos drama, to the tune of, *Let me take you on a sea cruise.* The Pax loved it but the Greeks didn't. There was a lot of financial drama with this ship. The gangway was often at half-mast while officials waited for payment. Jannie called us all together before we set sail on the last leg of the voyage and informed us that salaries *may* be delivered in Mombasa. Terry Fortune had had enough and disembarked in Durban. I excused myself from the meeting saying, "Excuse me, but there are hundreds of people who would like to go to Greece, and I am due on stage for the Welcome Show! Passengers are coming aboard; I'm going to get ready!" Everybody got up to get ready and Jannie gave me a wink. We set sail with about 80 passengers and Nadia's mom. Like mother, like daughter! She closed off a lounge for a few days with the passengers and no staff or entertainer was allowed in. She directed and choreographed a show in which the passengers took the Mickey out of all of us. It was wonderful.

Neil Wesselo flew to Mombasa to inform us that no salaries were forthcoming! (I don't know about the others, but I was never paid by Con Leach for the last few weeks on the ship) A lot of passengers disembarked and about 20 were still going to Egypt, along with Nadia's mom. Thank goodness she took the remaining pax under her wing and sorted them all out in Cairo. It was a ghost ship that sailed on to Piraeus. The Captain gave us all drinks in the bar and thanked us for enduring the hardships (?) and they moved all the stock up to Tom and Kiriyakos' pub. Kiriyakos gave me the key to the bar and his phone number in Piraeus and informed me about his bar *La Vita*, opposite the yacht club. I thought... *a ships barman owns a pub in Piraeus?*

The ship duly arrived in Piraeus and we weren't allowed off because we had no return air tickets. After a bit of an attack from Jannie and Nadia we found ourselves in a four star hotel in Athens. Tom flew to England after a few days under his own steam to visit his family, and we explored Athens and Pyraos. Ann Powers and Tam Minter found a lovely English Pub called Mike's Bar. It was our evening meeting place. We also experienced the heaviest snowfall that Athens had had in 40 years! Moss took photographs of me juggling snowballs at Parthenon and that evening had us belly dancing in the night clubs.

The next day I told the gang about Kiriyakos and his pub. They were all dubious but followed anyway. We found *La Vita* - Italian for *The Life*, and asked the lady at the counter for Kiriyakos. She said no and I pretended to strangle her! Right a way she was on the phone and we thought the police would be here any minute. It was all Greek to me as to what she was saying on the telephone, but she looked at me and said, “Robin?” I took the phone and Kiriyakos said he would be there shortly. When he arrived he took great delight in introducing us to all his family and clientele.

We sang and joked, and between the magic for his guests, we ate and drank way into the night. Lindsay, Sean and I were laughing on the floor when the sobering words of, *"Let's sort out the bill"* rang the air. We all piled our Drachmas together and Kiriyakos was annoyed. He looked at me and said, "*I invited you!"* and pushed all the money back! Lindsay wanted to see more of Greece, and Kiriyakos and his family looked after her for quite a while... all this from a guy who was introduced to me as the aggravating barman! *"Be Nice"!*

The newspapers back home were having a field day because the Oceanos *hero's* were stranded in Greece. What a great place to be *stranded!*

We visited SAA offices quite a few times and Geraldine had gone with Lindsay to the Eperotiki offices looking for work. They saw Captain Jannis Avranis with Takis, the chief engineer, both from the Oceanos. When they saw Geraldine, they ducked into a door and disappeared! They had been found guilty of crew negligence and they and their senior officers had lost their jobs (temporally so it seemed). I know they don't celebrate the 4th August like we do.

Our tickets finally arrived and after a farewell party at Mikes Bar with all the gang, we decided to call it a night when Ann had to carry Tam home! It was so nice to see an SAA Jumbo on the runway, and the first Castle didn't touch sides. I was off to Sun City and a whole new world!

SUN CITY - ESCAPADES - EXTRAVAGANZA

During a farewell party with Tracy Miller, from Starlight Cruises, and her friends, one of them raised his glass and said, "Your life will never be the same again!" He was right!

During the first day of rehearsals, I asked Dee and Roo (Wild Coast) who the girl was standing behind me during the finale bow. Julie Armstrong! I was in love and there wasn't a life boat in sight. She had a little girl called Bianca and I thought she was married, because Bianca's dad would visit every weekend. Deanne soon cleared that up. She was single!

Kim Dundas and her mom and dad were there from Thaba'n Chu- and Carol, my Achille girlfriend was dancing in the bingo lounge, and a host of new faces. I was introduced to an Argentinean called Normando Rojas. He was the other specialty act and he did hand shadows. We became great mates and he went everywhere with me... Joy Dundas used to sing, '*Me and My Shadow*' whenever she saw us together. We knew our act, so we had free time and I took Normando down to Johannesburg for a weekend.

The dancers and the head liners, Franco and Mary-Jane from the USA, continued with the rehearsals. I took Normando to a pub called Foxy's and we shared a table in this busy place with a few people. Abbott and Crab (Dave & Henry) were the band and are also great pals. They introduced me from the bandstand as the Oceanos chap and our table suddenly turned into a party. They discovered I was a magician and after a few tricks they asked Normando what he did. He twisted a lamp from the wall and had the place in an uproar with animals and shadow faces blazoned on the wall. We didn't spend much of our own money that day!

Nikki (Oceanos and Grahams flat) came to visit and also Brenda, a passenger girl friend from the Achille and Aegean Dolphin. But I had my eye on Julie. I did enough spade work to build the Lost City on my own! It was working! One evening while I was onstage she hid my denims. When they were eventually returned I told her if she wanted to get into my jeans, she should just ask. She asked me to baby sit one day while she was in rehearsals, and Bianca and I hit it off. I took this little one and a half year old around the complex, changed her nappy in the golf club, and went for lunch. Julie found us much later and laughed out loud - we were both covered in tomato soup and pasta! The next day Julie phoned my room saying that Bianca was driving her nuts! She wanted "*RB,*" so I joined them at Water World, took them for lunch at Kwa Maritane, and we had a lovely day together.

I was invited for dinner one evening and Julie was breast feeding when I got there. The show we were in was a topless show, and if you've seen one, you've seen them both! I said I'd like to try that but I didn't think that Bianca would like my boobs! We had a great evening and the next day I moved in with them, leaving my hotel room for friends and visitors.

There were loads of friends that came to visit, even Vaughan Leader and his daughter, Melanie, from Cape Town. I made Vaughan stand up to take a bow as the magician who helped start my career on the right path. Rupert and Cheryl Mellor came round, Clive and Jane Levy, my mom and Vic and also Tony and Nicola King. Tony has always been extremely encouraging and they loved the show. They also loved Julie and Bianca. Julie's mom and dad came over from England, and June and Pete met my mom and Vic over dinner. When they returned to England, Pete asked me to look after his girls - he could tell!

Moss and Tracy came to play in The Crazy Monkey Bar and it was wonderful being there with them. The 4th August was coming up and a coach load of survivors came to be with us at Sun City. I took my bow that night in my white ship's uniform and from the theatre we all went up to Moss and Tracy in their bar. I joined them on stage and together we did the Oceanos song and it turned into a very late night. The next morning we and the passengers had breakfast together and hugs and tears were the order of the day. It was so nice of them to make the trip up to Sun City just to be with us.

What with all the golf with Billy Domingo and the casino staff, Kevin Hollis and camping out in The Pilanesberg Game Reserve, life at Sun City went by far too quickly. The Achille Lauro was coming back to South African waters and Starlight Cruises wanted me on. Julie knew about the contract and it was very sad saying goodbye to both of them. I promised I would be back and also promised to visit her mom and dad in England at the end of the voyage. Then it was off to Johannesburg, to find those familiar faces waiting at the airport.

WELCOME BACK - ACHILLE LAURO

She had been painted a darker blue, but there was no mistaking her as we flew over Genoa. A new band joined us, Blend. Dave and Pete were to become life long friends. Most of the old officers were on board and they had read and seen everything on TV regarding the Oceanos. There was a new puppeteer on board, and he had a wonderful routine with a clown that I was told about and I shouldn't miss it. I was standing at the back of the room with Captain Orsi and Geraldine, and she said, "Watch this *Bokkie,"* (my nickname) and I watched as Steve Wright brought out this little clown. It blew up a balloon and flung his leg over it and floated about to the tune of *Send in the Clowns.* When he landed, the balloon burst, and this poor little clown sobbed against Steve's leg, with his little shoulders shuddering. Just then Bronwyn, the child hostess, appeared from the wings and tapped the little clown on his shoulder. She was carrying a new balloon and the clown gave her a big hug and she took him to the playroom! The tears were rolling down my cheeks and Captain Orsi looked at me and said, *"Some hero you are!"*

Moss and Tracy came on for a break from Sun City and I had decorated the bandstand with trees and teddy bear look-a-like monkeys to resemble the Crazy Monkey Bar at Sun City. They also had tears in their eyes when they came back from Steve's act, but they laughed when they saw their stage.

With the likes of Dave and Pete from the band Blend, the occasional visit from Ian and Jackie Sinclair, I'm sure we had more fun than the passengers. Ian's on board lectures were always entertaining. Titles like *Sex life on the Reef* and *The aerodynamics of a Penguin,* it was just as funny as watching him trying to sober up at 6:00am to do the dolphin and whale spotting. I don't know how many times he was asked as to what times do the dolphins come out! I'm sure he was asleep most of the time behind those Polaroid's. Other celebrities included John Berks, Jeremy Mansfield, Arnold Geerdts and Tony Factor came on for his honeymoon. You don't want to go out for drinks with Jeremy in Johannesburg - it takes too long to get to the bar - everybody stops him to talk. It's a bit like walking around with Frank Sinatra, I would imagine.

Dave went to his cabin one night to find a scene like *Psycho* in his shower. Pete had somehow lost his balance with the roll of the ship and fell on to the toilet brush… it went right up his backside. It was a very uncomfortable keyboard player we had for the rest of the voyage, what with his teeth and arse problems, but he visited us in South Africa recently and I was pleased to see his sense of humour had been surgically replaced, along with his arse!

Clem Leach, you've already met (fridge magnet) was also on during this voyage. On one of the cruises he was given the duty of looking after Mrs. Naidoo. Mrs. Naidoo was in a large wheelchair that couldn't really fit through most of our doors, and the ships wheelchair was a lot lighter. Clem was designated to look after Mrs. Naidoo in our Company Car. It was a familiar sight to see Clem, Mrs. Naidoo and the company car, travelling about the ship. Clem used to take her to dinners, shows, and the pool deck and also for evening drinks. One day it was raining and Clem had been for a bit of a nap-attack. When he came through the bingo lounge while I was doing the calling, I said, "Hi Clem, where is Mrs. Naidoo?" He raced out of the lounge to huge laughter, because all we could hear was, "*Aaaaaarrgh the pool-deck!*" Another evening south of

Madagascar, we were riding a tail of a cyclone. Clem brought Mrs. Naidoo into the Sorrento Lounge, parked her and the company car at a table, and went to the bar to order drinks. During the ships roll, Giorgio used to often place your drinks away from you, and tell you to wait. With the next roll, your drinks would come sliding towards you, and you would catch it on the bar counter. Clem nearly caught his order when he suddenly remembered! “Mrs. Naidoo!” he shouted, as he swung around and chased after the company car, which was gathering momentum through the chairs and tables, along with Mrs. Naidoo.

After our arrival in Genoa we all went our merry way. I visited Julie’s parents in Leeds, met loads of the family and friends in the *“Woodcock”* and joined Skip Cole in Bradford at the “Swing Gate” for curry and beers and left for home.

Julie and Bianca moved in with me after she left Sun City and together, we went to The Venda Sun for a show. We had a great time with their GM, Rod Walker, and with Bianca sharing our room. We must have watched *Peter Pan, Cinderella* and *Beauty and the Beast* about a thousand times. Loads of shows in the country during the rest of the year, but we were getting ready for the ship.

Back in Johannesburg, Starlight staff was getting ready for the next season. They agreed to have Julie and Beebs on the ship. We phoned June and Pete in England and MSC agreed to let them stay on board in Genoa, while the ship was getting ready.

TOMS VOYAGE

We all met as usual at the airport and Julie and Beebs were introduced to everyone. Nadia was Cruise Director and Geraldine, Lynne, Jane and Lindsay were on along with Neil, Sean and a few newcomers, Kevin Osborne, (son of Daphne of Starlight), Lee Foggitt (son of CJ-Goony Foggitt) Dale and the old mover himself, Clem. Also on board we had Roger Chucklefoot, Julian's brother. We did the old Mile High Club thing and arrived in Genoa the next morning. We found the ship in dry dock and this was Julie's first cruise. Lack of working toilets and running water, with a two year old on board! Moss and Tracy looked after Beebs while we went to the airport to pick up Julie's parents. We brought them back to the ship and introduced them to Tom and Terry Lester and the rest of the team. We had drinks that night in Tom's cabin and we were all geared up for the season ahead.

The next day the dry dock was filled and we sailed across the harbour to the passenger terminal. June and Pete left to go back to England and the ship was getting ready to sail. Beebs was wrapped up in her pink ski suit and very excited. The new passengers came on with a load of old faces. The entire Tote Auction crowd - it was going to be a fun filled south bound. I was sporting a shiner (Julie belted me for not telling her that Lindsay would be on board - warning sign one ignored!) Bianca was introduced as the youngest member of staff and the night we met the Captain, she was not keen at all. Peter Pan and the dreadful Captain Hook came to mind. She got over her fear and settled into the ship board activities. She became the ships mascot and was spoiled rotten. She used to perform with Van in the staff show, the oldest and the youngest, doing '*Animal Crackers in My Soup.*'

One evening Terry Fortune decided to do a mini cabaret show in the top Capri Bar. He was true to form and had the audience eating out of his hand. He got a bit carried away during a rendition by Shirley Bassey. He lay on top of the piano and when he put his elbow out to rest his head on, he misjudged the curve of the piano, and fell arse over kettle to the floor, and finished the song off from there! To thunderous applause, he took his bow, lying laughing on the floor.

Between Israel and Port Said, we lost a dear friend in the form of Dave, Allan Foggitt's father-in-law. Allan had Nadia ask me to accompany Dave and the authorities ashore to the morgue, being a friend to the family back home. Julie didn't think cruising involved things like this, but Tom & Terry Lester stayed up with her until I returned to the ship. I went by boat around the dark Port Said, to a waiting ambulance, through the dark streets and signed Dave in at the morgue. After identifying Dave and having to check for any jewellery, etc. they took me back to the ship. Tom, Terry and Julie were waiting for me in the Sorrento Lounge with Giorgio, the barman, and a few others. Tom said that if ever it was his turn to go, he would like it to be at sea with his mates to look after him. The Captain, who also had to wait for me, then took The Achille to its place in the convoy, and we eventually headed down the Suez Canal, into the Red Sea.

Julie took to the activities and Tom, Jane, Terry and I did the Auction every morning. Terry was selling the tickets in the foyer, when I walked past and he told me that Tom wasn't well. He had spent the night in the ships hospital on the nebuliser and was in his cabin. I went to his cabin and he was breathing very badly, but he said he

would be fine. He just needed some rest and could I give his apologies to the team and do the Auction for him?

I did the Auction that morning, and went to the bridge to do the Noon Report. Captain Orsi asked after Tom and I said I would see him in a few minutes when I took the Auction money down to him. I was relieved when I got there to find the cabin bare. I thought he was feeling better and had gone down to lunch. I put the money bag under his bed in its usual place, but when I stood up, I saw the reflection in the mirror to the open toilet door and Tom lying on the floor inside. I ran to him and tried to make him comfortable. I pumped his chest as Mario had taught me to in Fremantle all those years ago and gave him mouth to mouth. It wasn't working! I ran to the nearest phone and called the bridge. The officers and the doctor were there in a flash, but we had lost our Dear Tom.

I called all the available cruise staff and together we lay him on his bed and closed his cabin door. My cabin was across the corridor and bit by bit all the staff and entertainers gathered. Jane had been with us from the days of the Astor - when she walked into my cabin we both burst into tears. Julie took me out on deck with my sunglasses on and we just stood and looked out to sea. What an introduction she was having to her first sea cruise.
Nadia and Geraldine came to find us. I gave them all the cabin numbers and names of the passengers that would have to be told in the privacy of their own cabins. Tom was like family to all of them - these were the same people who had flown him to Walvis Bay a few years ago. Messages of condolences immediately started coming in from Starlight. We used to be inseparable, but we were now. It was 30th November 1993.

Back in the Sorrento Bar, Giorgio poured us drinks, came round the bar and hugged me, without being told, he knew. Bit by bit more sunglasses filled the room, until the entire Auction Team was there. Before bingo started in the lounge next door, brave Nadia stood up in front of all the assembled passengers, and announced in a very lump in the throat sort of voice, (she was also an Astor Girl) that we had lost our Father of the Seven Seas. Garth Victor and the band played Eric Clapton's *Tears in Heaven!* I don't know how Lionel Bastos ever got through that son I know the Auction Team and I couldn't listen to it for a long, long time without shedding a tear for our Tom Hine.

Nadia gave me the next day off and I didn't take part in the Crossing of the Line Ceremony. Terry braved the tears, as the real pro that he was, and he was King Neptune. I had to take hold of Bianca when she got mad with the antics on the pool deck when her mom was forced to kiss the fish and all the madness that goes along with the Crossing of the Equator by sea for the first time. The next day we arrived in the Seychelles, and Tom was taken off and later flown to England. It was in accordance with his will and he was buried by his family back home. This was the 2nd Dec 1993 - these dates will crop up later in the story.

When we left the Seychelles, Pat and George and a few others told me to go on with the Auction. They all agreed that Tom would have wanted it that way. So we did and a few days later, on the morning of the 6th Dec, before I went up to do the Noon Report, I announced that, with their permission, I would use the 10% of the day's takings that normally goes into the Golden Pool for the last day, to buy them all drink - *Today is Tom's birthday."* I don't think any of us went to lunch that day.

Christmas and New Year on board is without a doubt the best time for cruising. Family orientated for the Christmas cruise and all fall down for the New Year cruise. Allan was on board and gave us a wonderful speech of congratulations derived from the comment cards received from the previous passengers.

Bianca's Birthday was next, 9th Jan. and she was turning three. In the staff dining room, Julie had made a lovely throne for her, and all the gang made such a fuss. The strolling Minstrels came in singing Happy Birthday with a giant cake from the galley, and she took a long time in opening her presents.

On my birthday during a performance by Moss & Tracy, Julie arrived wearing a suit of mine and did a little rhyme, with magic and quips, agreeing to marry me. We had a lovely party that night and Bianca went to sleep with Moss and Tracy. My mom and Vic joined us during a Bazaruto cruise for our engagement party. The Captain and the Chief Purser, Senor Aldo Accardi, picked up the booze and the snack bill! Even John and Joan Foggitt. Snr were there. We phoned June and Pete and wedding arrangements were under way.

Bianca at the Helm

During the voyage, at a stop over in the Seychelles, we were all around the pool at the Coral Strand Hotel. Julie had gone to the bar and I was watching over Bianca. She came out of the pool and handed me her water wings and suddenly ran back to the pool and jumped in. I knocked our drinks off the table in my haste to get to her and when I surfaced along side, she spluttered, smiled and said; *"RB, I'm swimming!"* I took her out of the pool and said she must do that again when mommies back. Why should I be the only one to have a heart attack! When Julie arrived back I told her to watch this- and Bianca was off. *"Water wings,"* Julie screamed, and I dived in next to Bianca. All the staff cheered for their little mascot – she was only three. After that they both went up in tandem on a parasail ride, and a water baby was born.

During all of this, Julian and his girl JJ (now his wife), had spent the best part of two hours trying to break open a coconut. Eventually one of the locals came to his rescue. He shoved a sharp stick into the ground, whacked the coconut onto it and split the coconut in two. Julian exclaimed in Tommy Cooper's voice, "Just like that!" Mel Miller, the Biltong & Potroast comedian was on board. During a stop over in the Commores he was amazed to see the locals building a shelter, "They're knitting a house" he said.

Tony and Nicola King joined us for the northbound cruise and their children were christened in the ships chapel. I am now God Father to Kimberly and Jason. They had always wanted to see the Pyramids, so I phoned Madame Fatma in Egypt and organised a tour for the entertainers at a wonderful discount. We had a great time and during lunch at Giza with the Sphinx as our backdrop, our passengers came scurrying past on their tour in a flurry of dust. Imagine their faces while we were being wined and dined in luxury.

On arrival in Genoa, we had time to kill before our evening flight was due. While crossing the busy four lane road outside the harbour, I turned around to look at a car that was hooting frantically. It was Captain de Rosa. He piled Julie, Bianca and I into the car and took us to a lovely sea food spot for lunch. It was packed, but not packed enough to fit in the Heroic Captain. It was like going out with Frank Sinatra again! Imagine the faces of the crew when we pulled up next to the gangway and saw their old Captain was my chauffeur.

June and Pete had done a wonderful job in organising everything. The church was very old (1435 or thereabouts was the date on the stone at the door) and they had a power failure. But there was a pump organ and Martin Clifford and Pete Healey took turns in getting the music out of the pipes.

Tony de King was my best man, Terry Lester was MC and there were a host of entertainers and cruise staff that made it a wonderful party.

We decided to settle in Cape Town and performed in Follies Panache for Carlo Spetto - Judy Page and Sam Marais were also headliners in the show. Gerry Bosman did all the music and what a talent he was. I had great write-ups for the show, but Ricky Grey and I were not mentioned in the CAPAB posters. They advertised the producer, Sam & Judy, and the sets were built by Keith Anderson! At a party at Gerry's place, it was announced by the big shot of CAPAB (Come and Prance a Bout!) that the show was being extended, due to popular demand,*"Can you do something different for the rest of the show, Robin?" "No, nobody knows I'm in it, ask Keith to build you a new f***ing wardrobe!"* Gerry had just taken a sip of his drink during this conversation, and it sprayed all over his lounge as he laughed and choked.

Then it was time to join the Achille Lauro again, but Julie decided to send Bianca to her father instead of her coming aboard. Thank goodness!

ACHILLES LAST VOYAGE
(How can we sleep while our beds are burning?)

There were the old faces and a few new ones. Another welcome face was Thinus Maree (*Soft Shoes - Just for Kicks)* We did the usual cabin allocations and Nadia put us into Martin Clifford's old cabin, right at the stern door leading to the back deck. Captain Orsi announced we were to do a life boat drill in the harbour and an outside uniformed group came to monitor the whole event. (?) This was the first time I had ever experienced a drill in Genoa harbour. We went to our muster stations armed with our life jackets and Moss and a few others were lowered down with one of the life boats. They had to go around to the quay side to get off because they couldn't get the life boat back up again! The crew eventually did and we failed the drill.

The next day the uniformed group came back, but this time, we as entertainers and Starlight Staff weren't invited to participate. The ship passed this time and all the uniformed group were smiling as they left the ship with their duty free bags clinking ashore. (?) Senor Aponte stayed right to the end and watched the Achille sail out in the cold. Normally his limo would have whisked him off as he stepped off the gangway.

He listened to *La Nave Blu* (The Ships song) for the last time (?) and the *Symphony* hooted a farewell as we sailed out of the harbour. Julie said, "*I've got a bad feeling about all of this*".

The tote auction was under way and the shows were up and running. Everything was going smoothly and one night while the band were on deck a large shadow with lights loomed over the ship. An oil tanker passed by with what seemed like inches to spare. We nearly collided with the vessel. In the morning I asked Giuseppe Balzano about the incident and he said the tanker changed direction at the last minute. It wasn't written up in the log either! (?)

One morning I took two groups up to the bridge before the Tote Auction began. While pointing out a few Navigational bits of interest, I noticed the rev-counters were both in the red. Normally the Achille would cruise at 16-18 knots and the revs would be about 85/90 rpm. I asked Giuseppe Balzano the reason we were pumping the engines so much and he gave me the Italian shrug with both hands held up with fingers joined - you know the one.

The Captains Estimate that morning for the benefit of the Auction was about 380 miles. We did the auction and I went to the bridge to do the noon report and read the mileage out over the Tannoy, it was about 475 miles. The auction was way out and I told the passengers concerned that I had made a mistake in the morning and adjusted the chart and everyone was happy, except me. Normally I would also announce during the noon report about passing the Horn of Africa and *an infamous Italian* had built a lighthouse on the point. (I was never allowed to mention Mussolini on the ship) We had passed that point at about 9.30 that morning. (?)

Julie had collected loads of photos of Tom and put it all in a collage. She also made a beautiful invite to all of Toms mates, staff and officers to attend a memorial wake for tomorrow, 30th Nov 1994. I recall the cover had the music from Billy Joel's *Piano Man* and Chief Purser Aldo Accardi would donate the drinks and eats after the show for Tom's Memorial.

That night Julian, JJ and I did the Magic Night and after the second sitting show we all went into the Sorrento Bar for drinks with the passengers. It was just after midnight and a passenger came through from the Scarrabeo Lounge and asked Julian and I if we had left our smoke machine on. He said that he didn't even use the smoke machine. I went into the lounge and there was a lot of smoke. Suddenly "Jane's" fire door suddenly slammed shut! I thought, "*Fuck it, here we go again!*"

Passengers came out and everybody went out to the decks. George Gardiner went through every cabin on his Lido deck to get all the passengers out. (No alarms yet?) An elderly Dutch fellow had a heart attack and the doctor was trying to get him round. Julie was consoling the wife who was with me in the morning doing the bridge visit and I was chasing some people away with a video camera. He had obviously learned that CNN and SKY News pay more than insurance companies.

All through the night we went around talking to passengers, keeping them calm and all the while the guy's below were trying to fight the fire which had broken out in the engine room. In the wee hours we managed to end up on the pool deck with some of the guy's singing, "*How can we sleep while our beds are burning?"* One of our firemen (Niello) came past and told me the fire was nearly under control and this was also confirmed by two Italian mates of mine from the casino. The ship by now was starting to list to the port side, which would make launching life boats on the starboard side a bit of a difficult task.

Nadia called a meeting with the staff in a very smoky Arrazzi lounge and said that Captain Orsi was going to abandon ship. Outside on the pool deck some idiots had formed a human chain with water buckets and were throwing it down an air vent. I used to do engine room visits; this was like using an eye dropper down your air vent on the dashboard of you car while your engine is on fire. The Captain, in Italian, and Nadia in English, announced from the bridge that we were abandoning ship and nobody should go to their cabins due to possible smoke inhalation. Julie and I went to our cabin - we were right at the end of the corridor and I knew we would be alright. We got two jackets and sun block and she took the two photos of Bianca off the cabin wall. We left all our belongings behind, Christmas presents etc. and closed the door. Just down our corridor I heard voices and called out to them. Their response was, *"Hello Robin, we won't be long."* They were two bridge officers and they were busy opening an air vent? (P deck aft Port side) I remembered Julie's words as we sailed out of Genoa, *"I've got a bad feeling about all of this!"*

Smoke was pouring out from all over back on the pool deck. Nadia came past and handed me a bag containing entertainers and staff passports. I realised then that the pursers dept. (like the other ship) had obviously done their emergency drill bit, by going through the safe custody boxes and passports.

All of the Starlight team did a wonderful job of controlling the passengers and getting all and sundry into life boats. I remember seeing Kevin Osborne, leading an elderly gent through the smoke, holding a wet towel over the fellows face. One of the lifeboats wouldn't go down and everybody had to get off. The flames were now coming out through parts of the ships side and it looked like we were running out of time.

With all the other boats down, we now had to launch the life rafts. Unlike all other cruise ships, whose rafts are on a slide, quick release system, our life rafts were covered and tied to the deck. (Remember we had just passed the drill test in Genoa - *the second time round*) They had to be opened and then physically lifted up over the deck rail and thrown over the side with the painter tied to the ship to prevent the thing from drifting away. I started cutting the containers open while some of our guys battled to lift the thing. They checked over the side and started lifting. It's a bit like lifting a small car and by the time I was opening the 6th life raft, I heard a scream. I looked round to find the look of sheer horror on their faces. In the time it had taken them to heave this heavy contraption up to the

deck rail, a life boat had come around to the stern to take extra passengers via the rope ladders. The heavy life raft had smashed right into the middle of the life boat and although the canopy and steering consul had taken most of the blow, it also broke a passenger's neck. Unfortunately it's a sight that lives with me to this day. (I have since spoken to the widow while doing a show at the Wild Coast and re-lived the time with her. She lives down in Port Edward and is a very courageous lady.)

Bit by bit we finally all clambered down the rope ladders and climbed into the rafts. The guys below all gave Julie the whistle as she came down in her short skirt and G-string. We battled with some of the elderly helping them down a swaying rope ladder, but finally there were no more desperate faces looming over the deck rail. I wanted to cast off as soon as possible. There were large blisters of paint bulging out of the ships hull caused by the intense heat from within. Doing bridge and engine room visits I knew we were 45m from the aft engines. (The length of the original prop shafts installed after the war. Her maiden voyage was in 1947) I also knew that we burned, on average, 90 tons of fuel per day, doing 18 knots. We still had seven days sailing days to Durban.

That's in excess of 630 tons of bunker fuel waiting to go up in flames. I passed the word around to tie the rafts together so we could communicate and keep the spirits up, and when we would eventually be rescued by whoever, it's easier to locate and doesn't require loads of stops to individual rafts.

We drifted for hours in those things! The enclosed rubber smell and the movement of the sea didn't agree with most. A bum over the side was the toilet; next to someone's head getting rid of other stuff from the other end. Eventually we saw a spotter plane flying overhead and Julian threw a flare-canister into the water. Unfortunately the bugger threw it up-wind, and all the orange smoke came bellowing through the life raft.

It was nearly dark by the time we were finally collected by one of the motorised life boats. An oil tanker was the first on the scene and nearly all of the survivors were piling onto the *Hawaiian King* - A ship that normally only has 24 crew suddenly had over 700 people to contend with. Lee Foggitt, Dale, Moss and Tracy ended up on another ship with a few other passengers & crew.

It was bad enough getting off the ship, but trying to get elderly people onto a gangway from a lifeboat, with a swell running, is something else. I had to stand behind Alan East (a regular old passenger. *Bugalugs* was his Nom de Plume in the tote auction) and jump with him onto the platform and try and keep him upright upon landing. It was nearly dark and we were greeted with cheers from our mates. George Gardiner told me that two of our officers were ready to come aboard and resume the search in the morning - He was having none of that. *"Julie and Robin are still out there, go back!"* I suppose his menacing face at the top of the gangway made them turn around. Thanks George!

During the course of the night other ships arrived on the scene, along with two American war ships. They immediately got into action and sent food and medical supplies via helicopter. Our Achille store crew, led by Michaelle, suddenly got into action, and shortly most of the union on board had something to eat and drink. (Neapolitans) Julie and I and the majority of the passengers slept out on the deck, while some of the elderly were housed and were accommodated in crew cabins by those generous crew of the Hawaain King.

The next morning, we (cruise staff & entertainers - seems to be a trend when you sink) handed out slices of bread and boiled eggs. Some tried to take two or three and shortly my box of eggs had finished. Boy can people whine over an egg. The chief cabin controller of the Achille, Carlo *Capo Loggie*, stopped me on my way up to the bridge. He complained about the conditions in which he had had to sleep and now he can't even find water to wash and shave! As if I have now suddenly been employed by the Hawaiian King as chief cabin steward! There were no officers showing their rank. It must be some sort of a tradition after sinking to take off your stripes or rank. Maybe it prevents the German passengers from f***ing you up! It's bad enough sailing with Germans under normal conditions, but Heaven forbid something should go wrong!

Up on the bridge was a hive of activity, other ships had assembled in the night and the Americans had sent a task force on board to help with the evacuation. Allan Foggitt was contacted in Johannesburg and he told me to tell the staff that their jobs were safe. MSC were sending the Symphony down to complete the Achilles' season. Wasn't that lucky... to have a spare ship doing nothing?

I joined our Captain and the staff captain out on the bridge wing. I heard from him that he had climbed down the rope ladder from the bridge at the very end. They were watching the Achille burn through binoculars and when he handed it to me the eye lens was wet... I glanced over slightly and he had tears in his eyes. He was a coin collector and I had found a 1947 South African (Half a Crown) coin (maiden voyage date) which I had given him on embarkation. He said he left the coin in his cabin. We stood in silence watching the Achille. I had spent over four and a half years of my life on that ship! I don't know how many years he had done. We were supposed to have had Tom's Memorial wake tonight - it was the 30th November 1994, a year after Tom's death.

Nadia had everything under control with the Americans, and bit by bit people were being evacuated to other ships. Julie, Dave Smith-Howell and I volunteered to help dish up whatever lunch the Hawaiian crew could muster. First in the queue were the Achille crew. I told them to let the passengers in first but they refused. We are all survivors was their excuse. Not even the "rank-less" staff captain could make them budge. To avoid a fight we just carried on and finally it was our turn to eat and leave.

I collected most of our regulars together (Pat & Lionel were here too - *Oceanos* passengers) and 47 of us were on our way to the USS Halyburton ffg-40. We said goodbye to the gallant Nadia, and the remaining few and went across in rubber ducks to the awaiting Frigate. Organisation was the order of the day. In no time at all we were given toiletries and T-shirts, overalls and a place to sleep. The crew had given up their bunks and would sleep out on deck or in the magazine. Dean Hattingh, one of our Starlight Staff, 'celebrated' his

21st birthday yesterday on the life raft... that's a birthday he'll never forget.

Julie and I with 21st Birthday Boy Dean

I changed out of my tuxedo and Julie, Dave and I joined the rest of the gang in the mess for a briefing by the Commander, Rob Reilly. He made us all feel welcome and explained that the crew had given up their bunks for us and told us a few of the rules that go along with living on a naval Man of War. It was wonderful and we all soon adapted to a different life-style. I first went up to his cabin and sent a list of all the survivors I had with me on board, next of kin, etc to the powers that be.

Julie became the adopted cruise director and Peggy Grinaker was the party animal of the ship. She would keep us and the crew up all night with tales and laughter. She said when a ship is abandoned we should also announce that passengers must collect their life jackets and false teeth as "*Half of these people couldn't eat!*" The ships helicopter kept flying out to check on the Achille. The ships Tannoy announced that the Achille had gone down... it was the 2nd December... the same day that Tom had left the ship a year ago, in the Seychelles, to be flown home to be buried.

Position N 7 & 29 min E 52 & 0 min Depth 5012 meters

We were heading for Djibouti and on the last day at sea the ship's crew had a barbecue/braai out on the helipad. It was the first braai I'd ever been to without a beer. The American ships doing active service only drink once a month. We also put on a show for the gallant crew and magic (from a borrowed deck of cards and sleight of hand, etc.) jokes and also *If I were not upon the sea…* Astor & Achille stuff. We had all signed a huge sheet of paper with a wonderful inscription about where we were and how we were saved, and to these young men that were venturing into an unknown future with the Middle East War. We wished them good luck and Godspeed, back to their own Country and families. As an entertainer, I often wonder how we often manage to make people laugh and cry. “Pathos” Well, we all did, and the next day we arrived in Djibouti.

On the flight deck USS Halyburton FFG 40

DEPARTMENT OF THE NAVY
USS HALYBURTON (FFG 40)
FLEET POST OFFICE
AA 34091 1495

04 December 1994

From: Commanding Officer, USS HALYBURTON (FFG-40)
To: Mr. Robin Boltman

Subj: **DESIGNATION AS HONORARY CREWMEMBER, USS HALYBURTON (FFG-40)**

1. From 02 December to 04 December, 1994, you embarked aboard United States Ship HALYBURTON (FFG 40).

2. Despite the arduous circumstances that brought you aboard, your patience and cheerfulness throughout have been inspirational. Enduring hazards beyond your control, you arrived on our brow with great enthusiasm and adapted to cramped quarters, borrowed clothes, and the personally demanding environment found aboard a U.S. Navy Man of War.

3. In recognition of your uncommon fortitude it is with great pleasure that I designate you an honorary member of USS HALYBURTON's ship's company. Well Done!

R.D. REILLY Jr.
Commander, United States Navy

The French Foreign Legion came to take us to the hotel. Some ships had gone south to Mombasa and South Africa, while some had gone north. The Germans from our sister ship, *The USS Gettysburg,* had gotten there first (seen the UK ad about Carling Black Label? I'm not making this up). Starlight had sent emergency things over like toiletries and clothing and the entertainers and Starlight Staff volunteered (again!) to help. We handed them out to the entire crowd and while I was away doing something else, I heard Julie screaming and a staff member called me. I went to the lounge and saw Julie in tears, with a large sunburned, socks and sandals, German standing in front of her. He accused her of stealing things that should be going to the passengers! He had seen her putting something in her top pocket of her Halyburton overall.

She removed the pictures of Bianca out of her top pocket, and I flew over three tables to fuck him up. The ships photographers and George Gardiner stopped me and they fucked him up instead! After that I stood on a table and announced to all the Germans that they were now on their own. None of our staff would be at their beck and call and from now on and we would also be treated as survivors. That night the hotel treated us to a poolside buffet and one of our German passengers jumped the queue with a stranger, "He is the German Ambassador" and I said, "I don't give a Fuck if its Adolf Hitler, get in the queue!"

We eventually got to bed after most of the passengers realised the worst was now over and with many hugs and kisses we all made for our beds. Some stayed on in the gardens wandering around thinking of how things might have been.

The next morning the French Foreign Legion transported us to the airport to await our flights. A few of the staff and entertainers posed for a photo outside the airport building and up we went to the embarkation lounge. Upstairs on our way to the airport bar I was met by the large, overweight, socks and sandals German, blocking our path. He said, "I do not want you to go home with a heavy heart, I am sorry and we," pointing to all the Luftwaffe around him, stood up and applauded, and he hugged me "and we are sorry too!" I felt a bit like Julie Andrews in the *Sound of Music.* I must have done something good!

The various planes started arriving to return the crew and survivors home to their various countries, and of course there were loads of hugs and kisses and exchanging of names and addresses. Starlight had a special discount for survivors and I used to often see ex-passengers on a long cruise.

Eventually the SAA plane, Limpopo arrived and we all embarked and headed for home. We had to refuel in Nairobi and Harare before landing in Johannesburg. Van and Monica's daughter had tranquillisers to fly to Genoa before the sailing, because she was afraid of flying. Now after the sinking she was only too keen to fly.

We arrived in Johannesburg and Bianca was there with her father to meet us, along with the press and TV interviewers etc. (We were told not to talk to the press - too late - half the guy's had sold their videos to CNN and Sky day's ago and I was interviewed on the flight home) I phoned Tony King and asked if he could put his refugee mates up for a night. He was at the airport like a shot, but Starlight had made arrangements for us anyway. We flew home to Cape Town the next day to be met by the family and friends, just like the last time... *Will you stop giving us heart attacks!* There must be an easier way to get into the news!

Starlight Cruises gave us tickets to fly to England at a discounted rate. We flew to England and needless to say Julie's folks and family were only to happy to see us and Bianca after what we had been through and we spent a lovely Christmas and New Year in Leeds, New Farnley.

The Hawaiian King called in to Cape Town and off loaded a few of the Achilles' lifeboats and Nadia; she had stayed on board for the rest of the trip. Ian Weinberg, head of the NSRI, called and asked me to do a fund raiser for them in the V & A Waterfront. Julie, Bianca and I sat in the life boat over a weekend and people would throw money into the boat. During a performance at the amphitheatre, Ian sent Bianca around with a collection boat. She raised more money in 20 minutes than we did in the lifeboat!

It's only when you're home and unpacking you start missing things. Besides losing all the Magic and props, there are clothes, suites, books on magic, music and my *Hout Bay Passport*! You can't just nip around the corner to replace Magic Props. I was lucky in the fact that a few magician mates realised this, and various props were kindly donated to me.

After our return we bought a house in Durbanville. It was a lovely place and we had some great times there. Cruise staff and other crew would often visit when the ship was in Cape Town. Kevin, Heather, Brian and Janine, our neighbour's, are still in touch to this day. They made that area buzz! I met Herman and Nicollette Fick; their children were at school with Beebs. We eventually ran a theatre supper club together called *The Grange.* Barry Hilton, Johannes Kerk-Orrell, Blackie Swart, Dowwe Dolla and a host of other Afrikaans talent performed there. It was fun while it lasted.

One of our Achille dancers had fallen in love with a Capetonian. Sharon contacted us and brought Gavin around. They became regulars and I eventually gave her away at her Cape Town wedding and Julie was her brides-maid, with Bianca as a flower-girl. They are currently expecting twins!

I received a phone call from Pat, Terry Lester's dad, to tell me that Terry was in hospital. He had bone marrow cancer and he feared the worst. I was due up in Jo'burg soon, but I was too late to say goodbye to my old friend.

I visited Pat a few day's after the funeral at his home, and we sat under the tree where his wife's (Terry's mom) ashes were sprinkled. Terry had left an open unfinished bottle of J&B behind, and between Pat and I, we finished the rest. We chatted into the wee hours and I know Terry was watching us... *"My Scotch!"* Thanks for all the memories, my friend.

One morning I received a phone call from Captain de Rosa, *"Am I going to see you while I'm in your city?"* I was down at his new ship, the *Italia Prima,* like a flash, and Julie, Bianca and I were wined and dined like royalty. Gerardo de Rosa had had a mild stroke and he was battling with his right hand. He asked me to help him with his cuff-links. His butler, Lorenzo and Staff Captain, Mr. "G" Sylvestro Gentilliomo were also on board (ex Achille Lauro).

His ship was in port for four days and we saw him every day. One evening after dinner and the show, we were in a pub on board and I asked the *Muso* to play Simon & Garfunkel's *El Condo Pasa* for the Captain. He nodded adding,"O*h, you know the Captain."* It was his song on the Welcome Nights during the Captain's Gala Night.

The Captain smiled and complimented me on my memory and I went to the singer later and asked him to play the Achille Lauro's song, *La Nave Blu.* He said it is not allowed to be played on board, but said, "*I can see this is not a normal meeting."* He played the song and two of the waiters, the barmen and I sang along. At the end the tears were rolling down our cheeks and Gerardo asked me to order the next round while he looked away. He said, "T*he staff should not see their Captain with tears in his eyes".* He also told me that when the *Italia Prima* did its Maiden Voyage, he and Cecilia went to bed after the festivities on the first night, and watched Sky News in their cabin. His old ship, the Achille Lauro was on fire and sinking! Cecilia had said, *"Your old ship is annoyed that you're not there to save her!"*

We invited him to see the sights of Cape Town, the Mountain, the Wine-lands, etc. but he had seen and done all of that. I invited him to our home on his last day, which he gladly accepted, and had a braai. He was over the moon. In my upstairs bar at home which was named by Terry Lester's dad Pat, *The Upper Deck,* he saw all the shipping memorabilia, including his famous picture framed on the wall. "You're more sentimental than me," he said. He taught me to play an Italian version of pool on our pool table and we toasted the ship and Tom with Tankards. It was a very moving farewell when I dropped him off outside his ship at the V & A Waterfront, with my car parked right at the gangway. I had a last drink with Captain Gerardo, Mr. G and Lorenzo and the Muso played *La Nava Blu,* and I drove home with fond memories of some wonderful people.

It was about this time in my life that my marriage was going wrong. Julie and I got divorced and it was a very sad time for me. Friends were very accommodating and people would phone at odd hours of the night, just to make sure I was OK. Friends!

I was now free to travel and do what I wanted and I visited the ship, The Rhapsody, when it was in port. Schalk and Christine were now Cruise Directors and they told Allen Foggitt I was free for a while. I was taken on and turned 40 on the Island of Bazaruto. I moved the Roaring Forties to just north of the Tropic of Capricorn and wasted no time in the Off shore drilling department. I sat at a restaurant with Schalk and Christine in the Seychelles and we phoned Danny Fisher and the girls on Schalk's satellite mobile, "They've taken me to a shit restaurant with no floors, no windows and only palm leaves as a roof!" He replied, "You're in the Seychelles, you mucky bastard!" Danny got the shits again.

It was a year later when Danny and I were on a cruise and I telephoned Seychelles that we would be arriving on Elvis' anniversary day. I would do the magic and Danny would do Elvis.

The Coral Strand was only too happy because the weather had been so bad and they didn't know what to do with their guests. We were booked, and Danny, Colette and I, celebrated the booking at the Banana Cocktail Bar in Mauritius. When we arrived in the port of Victoria ,Seychelles, Julian Russell and JJ (Oceanos & Achille - now Cruise Director), had arranged everything for the passengers. Then off we went to the Coral Strand Hotel and Danny and I booked in.

All the entertainers that weren't working that night all came along and joined in. Magic with English patter doesn't go down too well with foreigners and after an early warning call to Danny *Mr. Entertainment* Fisher... "*Start the car!*" I yelled over the mike... I called him on. Danny, as always, was brilliant. All the other entertainers joined in and the Coral Strand had a great show. After the show we went next door to the Beau Vallon Bay Hotel & Casino and we celebrated into the wee hours of the morning. On our way back from the hotel, we decided to go skinny dipping, but Danny being the most sober and without June, decided to watch our clothes. JJ was a bit tearful trying to get Julian out of the water at the high water mark, with his willy, splashing about in the sand. About 14 of us all slept in the same room that night and when Danny and I got paid the next day we mentioned the fact that we had over booked the room. All they said was make sure they all have breakfast and we look forward to seeing you again.

Dick B. Morton was also on board and he is a legend on the Natal Coast. For more than 20 years he has played at the Beach Hotel on the Golden Mile. He fitted in so well to cruising that you had to get into his lounge early enough to get a good seat. He could play just about any song the passengers requested, and in Durban, I always try to join Dick and Joyce for a drink, or just to visit.

The corporate market was very good to me and I did loads of shows for Bob Thornley and his production company (His brother-in-law Peter, along with his wife, Hilda, are now our neighbour's). I had done numerous shows for him in the past and on one occasion I was booked to do a Cadbury's Chocolate Function. The Fish River Sun is always a lovely venue, especially when Neville Williamson is around (Uncle Nev. as he was affectionately known during his days at the Royal Swazi Sun). We did numerous takes in filming for the Cadbury's Conference in shopping centres and the chocolate factory in Port Elizabeth, and eventually I was flown to East London to do the show.

I was to be collected by a guest speaker Laurie Kaye, who was the pilot that flew the Jumbo over Ellis Park during the Rugby '95 Rugby World Cup which we won by beating New Zealand with a drop goal by Joel Stransky. How will we know each other, he had asked Bob. "*He's a magician and you're a pilot, you'll find him.*" He arrived first and was waiting for me amongst loads of other people. I walked in and looked at him in the crowd and pretended to be flying, with my arms extended. He mimed to be pulling a rabbit out of a hat! We had met at long distance! During his show of the big moment, the closing music is *I believe I can fly...* you must see and hear it! We've been mates ever since.

Arnold Geerdts and I did Dave Callaghan's benefit night down in Port Elizabeth. What a great evening for a great sportsman and a gentleman! There were English, Australian, Kiwi's and a host of sporting celebrities. Guest speakers, Rupert (*Spook)* Hanley's painting's up for auction, it was a night that I'm sure Dave remembers as much as his batting knock of 169! Ian Healey from Oz is a wonderful guest speaker, and he apologised on behalf of

Sean Fitzpatrick, (remember Dave played cricket and rugby) who couldn't attend, but please tell Dave and the crowd this story...

After a match Sean was leaving Sydney from a hotel and he told the cab driver to take him to the airport. When you're famous (I believe) people look at you and you can see in their faces with that, is it him type of look? Where do I know him from? Eventually the cabbie looked at him in his rear view mirror and said, "Well, are you going to give me a clue?" To which Sean replied "I'm Sean Fitzpatrick, Captain of the All Blacks, etc. And the cabbie said, "No mate, I meant domestic or fucking international!" It brought the house down.

*I'm tired of trying to find the £%&**** things on the key board, so I'm just going to say fuck from now on, because you say it like that when you're reading anyway - OK?*

During the Millennium New Year Celebrations, Laurie Kay phoned and asked if I would accompany him to London on a trip. SAA Pilots could take their family on any trips, but this one was special - they were allowed to take friend. He had wanted a nutcase magician with him! Unfortunately I was already booked for a show in Windhoek, for an old friend who was running the Kalahari Sands, Tony Boucher. Years ago in Cape Town I took a client out from Nedbank, Cindi Crossley, (the one that got away) to the Newlands Hotel, and on arrival I realised I had left my wallet behind. I went to the duty manager (Tony Boucher) and told him about my predicament. "*Sign for every thing and come and pay me when you are next in the area."* How the world goes round...

HOW THE WORLD GOES ROUND

I was flown to L.A. in the USA to do an interview for the episode, *The Savage Seas.* What a treat! I was picked up by a small Mexican driving a stretched Limo. I picked up my bags, but he took them from me. I didn't think he could manage them, he was so little. When I saw the Limo in the car park, I immediately took out the video camera. (In those days for you younger readers, cell phones only Phoned!) I wasn't going to miss this. During the journey to the magnificent hotel in Pasadena, I lowered the tinted windows in the back. He offered to turn on the air-con and I said I wanted people to see me in this stretched limo. I could just imagine the questions on the people's faces as we drove by, "*Who the hell is that?*" The shows were all about upcoming TV specials and the Oceanos documentary was a big hit.

I had to do a questions and answers with the press and the local TV stations and also the guy's who had made me 'Personality of the Week' (an ABC TV Show). It was so well received that afterwards they took me to one of the lounges for drinks and I entertained them for a while. That evening I went to the Magic Castle in Hollywood, and only members are allowed on Saturday nights.

Chris Macan (one of our magicians and passenger you've met already) had emailed them about my arrival and reasons for being there. Open arms and I was wined and dined by the wonderful manager, staff and local magicians. Jet lag started taking its toll and I left about three in the morning. Man, what a trip and on the long flight back to Cape Town all I could think about was the rest of the team from the rescue operation. How they would have loved all of this. On all of my interviews I always talk about the team effort, which it was.

I eventually moved to Table View in Cape Town. Julie and Beebs lived close by and the visiting relatives from England still kept in touch. Pat and Tony used to stay with me when they came over. Now they're all permanent residents of South Africa.

In the area is a wonderful pub called the Royal Oak. My old muso mate Robbie Woodward and Linda own the pub. With Pete and Tim and all our local golfers of the Royal Oak Pub, we had some cracker times there.

Harold Body was there and I used to see his daughter, Carole, quite often. She was my first girlfriend when I was about eight years old. They lived below my Granny "B" in Saunders Rd. Bantry Bay, where I spent a lot of my school holidays. To bump into them after 20 years was just magic. Of course the visiting relatives were soon adopted and now they are locals! It was a time when I did a lot of cycling on the beach front and fishing because all my mates worked during the day. In the evenings we would meet at The Royal Oak or sometimes I would do a show somewhere and catch them later. I got involved with Tim and Mariaan with Traditional English Pork Pies, and spent a year with Tim delivering these pies to various places around the Cape. Mariaan now runs the pie business, so please buy our pies at any Spar, Pick 'n Pay or any of the other outlets - we need the money!

I did a few shows on various game farms and at Kapama Game Lodge; I spent 7 lovely days with the crowd from Isuzu Trucks. It was during this travelling conference I met Gerhard de Lange. An amazing motivational speaker using Lions as his medium and he was also responsible for looking after the lions when they were introduced into Pilansberg Game Reserve at Sun City. He lived out in the park and even the elephants came to know him. I was later on in life a guest at his wedding. I spent a lot of time on the game drives and learned quite a bit. One thing I did learn was that if you come face to face with a lion, you must maintain eye contact at all times. Take a step back and a step to the left (*eye contact*), then another step back, and one to the left - a step back (*eye contac*t) and another to the left... because you don't want to walk in your own shit!

I met Johann from Kapama who, along with his wife, started the Cheetah Rehabilitation program. They also head up the Endangered Species Protection Unit (ESPU).

On one occasion, Arnold Geerdts and I did the launch for the new Castle Light launch. Travelling around the country we had a real ball, and after breakfast with Graham in Durban, we were taken to the airport to fly to Cape Town. The naughty bug bit my arse when we sat in our seats as we were just in front of the hostess during the in flight safety drill. I told Arnold of my intentions, and he agreed. When the hostess pointed out the various escape exits, (you know how they do it, waving their arms about, pointing aft, middle and forward) Just then we jumped up and loudly sang a popular song that everyone does the actions to, *"Hey - Macarena!"*

Of course everybody recognised Arnold and laughed like hell, and when we disembarked, the air hostess with streaks of mascara down her cheeks said that she would not be able to do the drill again, without thinking of us. I don't think we would attempt that in this day and age.

After a cruise I went to the Grahamstown Festival to do a fund raiser for the NSRI. With the flying from Durban and the drive, I walked in as most of the audience were being seated. Except for two, *Boet & Swaer* were waiting in my dressing room with a six pack of beer. We had first worked together on the first televised *It's a funny Country* a few years back in Oudshoorn.

One evening I did the entertainment and auction at a black tie event for the ESPU at the Spier Wine Estate hotel in Stellenbosch. I invited my old flame, Cindi, (the one that got away) and it was a lovely function. I used a live cheetah in my act to choose a card, " *When I cheat, I use the real thing,"* I told them. Johan's pilot would hold up an item and I would take the bids. The guest list was something else including new and old Government Officials. I was pulling false bids out of the air for the ones I didn't like. Former President F.W. de Klerk was there and when he started bidding for something, the people didn't want to bid against him. "*Come on, he's not the President anymore"* and he stood up and replied, "*Yes, come on…I've always wanted some opposition!"* One of the items was a beautiful carving and he asked what it was made of. I felt it and replied, "*Rhino Horn!"* Laughter all round!

RMS ST. HELENA

I did a Box and Dine at Calvin Grove and also on board the SAS Tafelberg for Chris Armstrong, who is involved with the St.Helena shipping crowd and the Island itself. I met the Captain of the RMS St. Helena and some of the Islanders. The wheels went into motion and Chris soon got me involved with the upcoming celebrations for the Island's 500 year Celebration. I would sail out of Cape Town, spend a week on the Island performing and playing golf in the Bell's Sponsored St.Helena 500 Year Golf Challenge, and a week to sail home again. I love this job!

We had a send off in the Royal the night before and the next day in the famous Portuguese Embassy Pub and Restaurant, the Vasco da Gama. All the boys were there and I happened to be sitting near a guy who recognised my laugh. Turning around I found Gavin Durrel, one of the officers from the Astor! What a small world! (...but you wouldn't want to paint it!)

We had a corker of a lunch with Trinchados and Espetados (the auto spell check on my lap top suggests *Truncheons and Desperado's)* and a ton of Katembas. Katembas are lovely drinks served in a beer tankard consisting of Red Wine, (*Tassies)* and Coke. The boys were all fluent in Afrikaans by the time they drove me down to the ship, with Pete and Tim doing their impressions of Laurel and Hardy and 'Dinner for One' - *same procedure as last year, Sophie me love?"*

I was piped aboard and looked around to see who was behind me, but there was no-one. One of the officers came to me and requested me to accompany him to the bridge to meet the new Captain. Rodney was the first St.Helenian to Captain the ship. He shoved his hand out and said, *"I feel I've known you for ages. We use the video of the Oceanos as a crew training exercise!"*

Down in the lounge we got into party mode and another lovely addition to this trip was Allan Simmonds and his wife. (Astor/ Argus - you've met before) Allan would also be doing the Whiskey tasting, on board, on the Island and all the prizes at the golf, for Bell's!

I was given a lovely cabin up on the Bridge away from the passengers - I think Rodney wanted me nearby? I felt the nudging of the tugs in the early hours of the morning. I went out as the sun was just peering over the mountains. It was a perfect morning in Cape Town and the South Easter had been given the day off. I stood on the bridge wing and watched as we disembarked the pilot outside the harbour, and we were under way. We sailed inshore of Robben Island and I could see Table View and up the road, to a not yet awake Royal Oak Pub. After breakfast I went up to the pub for coffee (?) and saw my Knot Board proudly hanging on the landing, along with pictures of the Falklands, Royalty and

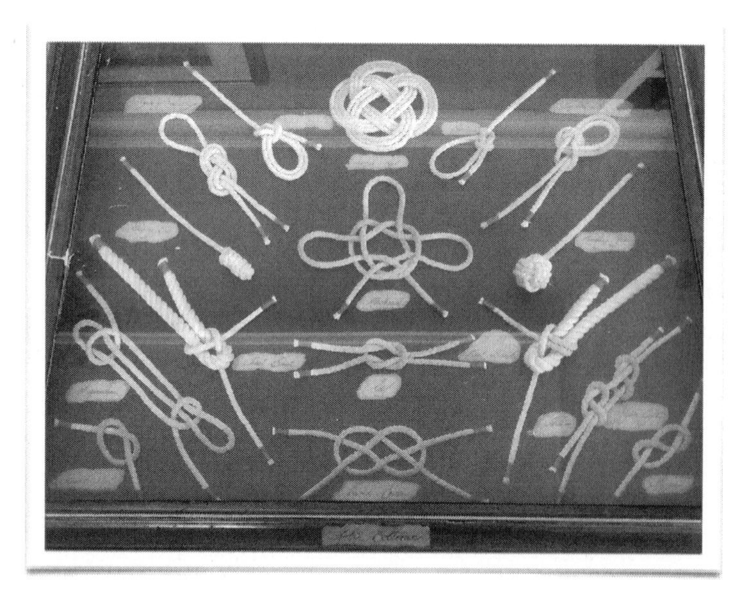

other things from the Island. (My Knot-Boards are a collection of a nautical knots used by sailors.)

Inside I was introduced to David, the Governor of the British Island we were heading for. We got on very well and whenever I see a Carling Black Label beer can, I think of Dave. When we eventually got off, I think they had surgically removed it from his clutches!

I did a few performances on board and also taught the crew some of the things we used to do on the previous ships. They also had their own entertainment and we all enjoyed cricket and ten pin bowling on the deck. After a week of sailing we saw St.Helena on the horizon. An imposing Island and I wondered what went through Napoleon's mind when he saw it for the first time.

It's a different sort of landing because the ship doesn't tie up alongside. They have a ferry boat and a huge pontoon for bigger things, like cars, etc. Remember the RMS is the life-line to the Island and its people, and I stand corrected, is also the last remaining Royal Mail Ship in action. Another method of disembarking passengers is the *Sky Lift,* and anybody that doesn't feel quite like clambering down the gangway in rough weather can use the Sky Lift. It is a metal cage that goes off the front deck on a sort of cable (like a Boson's Chair) onto the dock-side's floating pontoon.

Most of us stayed at the Consulate Hotel and I would also be performing at the hotel. My picture was all over the Island and wherever I went people came to introduce themselves. On my first day I decided to walk up Jacobs Ladder. It's 699 steps to the top and the view, when you get your breath back, is awesome. On the completion of the intimidating downward journey, I popped in to the nearest open door to find a chair. It was the museum. A lady brought me a glass of water and said, *"Just done the ladder?"* She had obviously seen the look of sheer terror and exhaustion of thousands of others before me! I eventually got to my feet and had a look around the place. Fascinating, and when I left she said, *"See you at the show tonight."*

There are also a few photos of one or two guys displayed at various watering holes. These are guys that have been banned for up to 6 months. If they got out of hand in a pub, they were banned from every drinking establishment (all six of them) on the island. There is also a fine if you bought them a drink. It is a very friendly place and I also met a budding magician called Robert. He phoned and asked if I could meet him when he had finished work for the day. He was a policeman on the Island and was in charge of the jail house. He had to wait for the two inmates to come back so he could lock them up for the night! *"Tuesday's they go fishing, and Thursdays they pick the coffee beans,"* he explained! Lovely, hey? (Maybe they were a branch of the ANC?)

I met with Robert and he locked me in Napoleons first cell for a photograph and later we discussed magic into the wee hours of the morning. Napoleon didn't stay in jail for long and the house he lived in was given to the French Government. He was later exhumed and buried on a different Island. Born on one, jailed and died on one and re-buried on another - a real Robinson Crusoe. The St. Helena Golf Course is notorious for being the most remote course in the world and between two continents, in the heart of the Atlantic. I met the golfers and we enjoyed the few days playing with them. Before tee-off one morning, I was asked to do a show at the local old age home. What they didn't tell me was it is also their mad house!

They had the tea and cakes all arranged and ready for after the show, and the Matron introduced me, and screams ensued from some guy who leaped from his chair and started running about. Everybody tried to catch him and calm him down. Eventually he was captured and he sat on some woman's lap for the duration of the show, trying to breast feed through the poor old dear's blouse. It was a bit distracting for me because the guy was about 25 years old! During the very short show my mind was thinking back to my mates at the Royal Oak… they would be thinking of me on this exotic Island with a long drink in my hand, etc. In my mind I was humming the song, *"If they could see me now, that little gang of mine"*

A different Rodney (golf captain) had dropped me off on the way to the golf course. It was nearby and I said I would walk back when I had finished the show. We arrived nearly simultaneously and he said, *"That was quick!"* I said, *"You didn't tell me it was also the fucking madhouse."* Laughing hysterically, he took me into the clubhouse for a much needed drink. When the game was due to start he announced, *"Jack Nicholson (Cuckoo's Nest) to the first tee,"* and I shanked my first shot out of bounds from laughing!

There were quite a few players in the golf tournament, locals, English, South Africans and a few guys from neighbouring Ascension Island. A local guy of about 70 years old, with only three clubs and a putter in his bag, won the tournament! Local knowledge wins again!

I had to turn down an invite to the Governors Residence for drinks and dinner that evening. It was the same time as the Golf Prize Giving, but the Governor was understanding and said he would be down at the waterfront the next day to see me off. More magic and drinks and group photo's later, we were driven back to the hotel. The bar was still open and the Governors party had waited for our return. Magic, drinks and more group photo's later; we went to our rooms to pack!

We had to go down to the harbour to join the RMS St. Helena. Dave, the Governor was there also, and we shared a few Black Labels together. (It had grown back, I thought!) Robert was also there and had sent his two inmates out to pick coffee beans, or something. One of the ships officers came up to us and saluted the Governor and said, *"Sorry Sir, but we need your drinking partner on board!"* After fond farewells, I wobbled off to the ship. Everybody was already on board and Rodney took the ship around the Island for a last look, before we headed SSE towards Cape Town. I had also been given a cabin nearer the lounge/bar! Yahoo!

The last few days on board were great. By now we were all mates. We played cricket again and I spent the last few days making new balls for the ship. They use *Monkey fists* - a woven, rolled up rope decorative knot used as a weight to throw a line. I also taught some of the pax and crew how to do it. I think the RMS St Helena has enough now until I go back one day. It's a holiday I would recommend to anybody with a passion for sailing and people. May the stars always guide you, Rodney & RMS, and smooth sailing.

It was our last morning on board, and we were just cruising into Table Bay. Tim called on the cell phone and was waiting on the beach front with the Pie Wagon, and I could see it on the beach front at Green Point. Whales and bird life are abundant just off shore, and I went to the bridge to warn the approaching pilot boat of a Southern Right Whale, that was alongside. We all met for drinks and group photos as we sailed into Table Bay. It was evening before we staggered into the Royal Oak and the reunion was, as usual in the Royal, a humdinger.

(The RMS is now sadly no more and the Island has since built an Airport. Cruise liners still call in on the Island.)

LIFE'S LIKE THAT!

I went up to Durban to do a show, and as always, contacted Graham. We met up afterwards and went for drinks. He was telling me that the Hotel he owned in Mooi River needed some help, in the form of a new GM, would I come and help? After a few days on my return home, I phoned Graham and said I that was keen on the job. I would be closer to Durban and Jo'burg and on the doorstep to the Drakensberg and the Midlands Meander, where many companies have their conferences, so it would also benefit my Magic world.

The farewell at the Royal was great and Robbie had agreed to take me for a last lunch at the Portuguese Embassy. George, the barman, and your on duty waitress *Psychiatric Nurse,* Annie, informed us, "*Yours is on the house, thanks for all the support!*" Robbie was really happy after seeing all the Katembas going down!

On the Sunday of my last day, the Royal Oak Golf Boys played the last game with me in Parow, and we all made our way to Tim's for my farewell braai. The boys all clubbed together (no pun intended) and bought me a huge, bright yellow driver golf club. When the guitar came out, we were all in fine voice. But the tears rolled when I tried to get through John Denver's *Leaving on a jet plane.*

I left Cape Town in the early hours of the morning and the cell phone kept me company, along with my worldly possessions, with all the guys phoning during the long 16 hour drive, en route to Mooi River, and a new life.

THE ARGYLE ARMS COUNTRY INN

The Argyle Arms Country Inn was glowing in the fading evening light as I made my way into the office, where Grahams daughter, Helen was waiting. We first went into the bar for a much needed beer. I met a few of the locals, Graham Goss, Teabags (Justin), Wayne and Mr.Plodd (Grant policeman) who all immediately helped me carry all the contents of my car into my new room and home.

I met the staff and Helen and Graham had planned a welcome Party for me on the weekend. Ian and Rose from Cape Town, who were at my farewell party, also attended my welcome party five days later in Mooi River! Most of the locals were there and everybody in this little community is ever so friendly. I was adopted and my first game of golf with the likes of Alen Butler, Bruce Bowley-Smith, David Boast, Mickey Taylor and Luke and others, was a great day.

I was fined for having golf clubs called *Magician,* (the brand really is a bargain from Fred Beaver at the Wild Coast) the brightest yellow driver they had ever seen and for advertising the hotels up-coming events! It was very handy being in the entertainment world and I knew all the pros. Tony and Nicola King, Barry Hilton, Dave (Achille Lauro) & Craig (Jester), Martino, Wolfgang Riebe and others all performed there. Some would be driving by and would pop in for lunch or to stay over. I know the locals enjoyed rubbing shoulders with celebrities, with the likes of Laurie Kay, Arnold and Lilly Geerdts, Wolfgang Riebe and The Bats, etc.

One day an elderly gentleman came in and asked why the piano was locked - I opened it up and he sat and played away; only then did I recognise the now aged Peter Maxwell." *What the fuck are you doing here,"* he asked? (He knew me from the shipping days and the magician from Cape Town)

Tom and I used to visit him often at the Edward Hotel in Durban when the Achille was in port. I promptly booked him for a show, and he hadn't lost his touch or lightning wit! (Michael Depinna's dad who was the GM there for many years gave Peter a 21st. Birthday cake - that's how many years he had worked seasonally at the Edward)

Dave Sharpe, Lloyd and Judy Emmanuel, (who own Zulu Falls, a must when visiting the area), Rolf and Kattie, Neil and Mandy, Wayne, Brian Memmell, who assisted me in editing this book… the list goes on and on. They all become like family in a town like this. It's a very small town, no traffic or traffic lights for that matter (two driving schools tho' and no traffic lights). All the books in the library have been coloured in. The story goes… *One day a guy came into the library and had had enough of life. Where are the books on suicide? The librarian told him where to find them, and it was quite awhile before he came back and said, "There are no books there on suicide!" She replied, "I know, they never bring them back! "*

On Tuesday evenings we held a Trivia night and some of the locals suggested an old tradition, *"Pig's Head!"* This is a roasted pigs head on a platter with *soldiers* of toast. It was duly delivered from the kitchen and Justin promptly stretched its jaws open and the locals dipped their slices of *soldiers* into its head. I now resembled the song, *a whiter shade of pale,* and Neil held out his hand to give me a gift. It was an eyeball, and it promptly flew out of my hand and around the bar! Eric - Bongani, the barman, just laughed!

Eric's girl was pregnant and when her water broke, he phoned me. (Why do babies always arrive in the middle of the night?) I drove them to the hospital and in due course, we went to fetch the new family addition from the Escort Hospital. As we were about to leave, a fellow came staggering up to the hotel *bakkie*, and Eric's girl screamed and said, *"He's a hijacker, he's got a gun!"* The fellow had his right arm under his jacket. I thought, not will I be hijacked! I flung the door open as he approached the window, and knocked him sideways. He screamed, and as I jumped out, he showed me his arm. Bearing in mind, we were outside a hospital; he wanted a lift to Mooi River, and his arm under his jacket was covered in a brand new plaster cast! Sorry about that! We gave him a lift.

Down in Nottingham Road one evening we surprised two old mates during their show - *Boet & Swaer!* (From the Oudshoorn Show) I did ten minutes with them in their show, followed by drinks at the *Bierfassel! It was the first time the Midlands* folk had seen me perform. Now my foot was firmly in the door and the hotel got a lot busier.

On my first Christmas and Grahams birthday, we all had a lovely lunch in the hotel. Fire roaring in the fire place (it gets cold there, even during the summer rains) and locals popping in for a Christmas drink, and to get out of helping to clear up the wrapping paper or help with the dishes at home! We had mistletoe hanging over the entrance to the pub area, when one of the local *Mooi River Open,* walked in and stood underneath, *"Will you kiss me under the mistletoe?"* I said, *"I wouldn't kiss you under anaesthetic!"*

The next day we were invited to Rolf and Kattie for a champagne breakfast, lunch & dinner, (well that's how long Graham and I overstayed our welcome) and met some more people, Robin and Denise (the Chemist) and a few others. The host, Rolf, took me into his Boma and we ended up playing the guitars until Kattie dragged us back to the party.

Ian was our Maintenance Manager and we got on like the old house on fire. After a great party on New Years' Eve, we were all recovering the next day. Locals were in and two weary travellers booked in - two sisters. I invited them to join us and one immediately claimed Ian's Water World cap. The young one was very tasty and when she stood up and announced she was going to bed, the whole party was a bit crest fallen (and other bits). But she looked at me and said, *"You will have to follow me!"* I looked around for a life boat and followed her. Happy New Year! I think I'm going to like it here!

Every January Graham and I used to go to sea on one of Starlight's Vessels. People in the area know of my past, regarding shipping. *Never go to sea with him, they all sink!*

A rep, Norman, calling in the area also used to frequent the ships. Robin the chemist asked Norman if he knew me from the ships because *he's at the hotel!* Storming Norman burst into the pub and a great reunion was had. I said we were off again for a short cruise and Norman promptly joined us.

Graham, Norman and I were in full swing on the pool deck before the Rhapsody sailed out. On board were some ex passengers from previous voyages (When they recognise me they would often say things after life-boat drill like, *"Where can we find you if the shit hits the fan"*), some regular crew members from the Achille days, some of the entertainers and cruise staff we knew, of course. The other surprise was Ian, Jackie Sinclair and family, Daryn and Kiera. Captain Orsi knew I was on board and our little party was immediately invited to the Captains Table for dinner. I'm not sure what the meal was like, but the gravy consisted of white wine, then red, followed by Moet & Chandon and Martin Clifford's favourite tipple, Remy Martin. The next morning some of the entertainers wanted to know how I enjoyed the Welcome Show last night. After dinner with the Captain we barely made it to the cabin let alone see the show! It's always great being at sea, and the people are truly amazing. I helped out one evening during the Mr. & Mrs. Quiz.

Julian & JJ (Cruise Director and soon to be married), were on board, with her parents from the USA, and I took the mike from Julian, as arranged by the staff, and they joined the panel." *I've changed the questions,"* I told him and the audience. That started the ball rolling, although I was on as a passenger, it was requested by the passengers, that I do a bit in one of the shows. Capt. Orsi sat at the bar watching with some of his officers and after the show, I was summoned to the bar. In his Italian accent he said, "*You like red wine?"* I said,*" No thanks; I've already got a beer."* To which he replied, "*No, you're like red wine, you just get better!"*

On a more recent trip, the Cruise Director was Stephen Cloete, He invited me onto one of the shows one evening, and as my name was announced, Ricky Grey (Magician & Illusionist from *Follies Panache*) called from backstage in a loud voice and said, "Remember Robin - don't say Fuck." Needless to say I walked on already laughing- and the passengers could hear him over my radio microphone from the dressing room.

We all had great fun, old passengers, entertainers and crew together after all this time. Of course, after a few beers, the memories started with all the funny things that used to go on. (Especially behind the scenes) *Remember the time when…*

A woman came up to us in the lounge after the show in her night-gown. *When was the band going to stop playing?* The music in her cabin was deafening. The band had already finished and I said to the guys, *come this will be good.* Her cabin was below the stage and I opened her cabin door and sure enough the noise from the *Band* was deafening. I reached behind her door and turned the in-house music wall radio off in her cabin!

Passengers strolling through the gallery of the ships photographs on display for the passengers to purchase… *How do we know which are ours?*

Does this lift go to the front of the ship? No - it just goes up and down!

Informing them after I was questioned as to the water in the pool (which isn't a bad question if you hadn't been in yet, and the Astor had both) as it was swishing backwards and forwards - *Is it fresh water or salt? Salt-water, every morning we pump new sea water into the pool.* She turned to her friend and said, "*Oh that explains the waves!*"

Standing on the beach with some first time travellers in Mauritius… *Wow, the time changes from leaving Durban, different currency, etc. - how high above sea level are we now?*

Does the ship generate its own electricity? No - we unravel an extension cable from the harbour when we set sail!

You don't see the crew around at night (unless you're lucky) - *Do the crew sleep on board?*

What time is the midnight buffet?

Ian Sinclair, like I mentioned some time back, used to do whale and dolphin spotting on the pool deck - *What time do we see them?* We all had a good laugh…and they're all true!

Robin & Barry Hilton

BACK TO WORK

Back at the Argyle we continued with day to day events. Danny, Wally and Ginger are also regulars, and Sunday was always a great day for me at the hotel. We would play pool in the morning and get ready for Karaoke in the evening. Everybody used to sing along and sometimes Des would also be there with his keyboards. Ginger would dance with any girl all night if he could and he is in his 80's! He and Kate (Helen's daughter) would often dance on the tables. I told Ginger one day that he should find a partner, "*What do you look for in a girl, Ginger?*" He replied, "*A pulse!*"

Danny Fisher's girls did their first show there and have since gone on to perform in the Far East. (Shanghai) Now the whole family (June, Joanne and Danielle) perform with Danny, and the Danielle I knew as a lump in June's tummy during my Wild Coast years, will have just turned 21 by the time this is printed! (The Baron von Trapp family singers - eat your heart out) That reminds me of Peter Taylor (a very versatile entertainer and friend) telling us a story one evening on the Achille...

A ship was sailing through the mist and the sailor in the crows nest shouted over to another ship, "Ahoy there, we are the Pequod, under command of Captain Ahab. We are in search of the great white whale, Moby Dick. Have you seen Moby Dick?" A voice came back saying, "No, but we've seen the Sound of Music about 7 times!"

In September one morning, David Boast, a regular, came in and started making the fire in the pub. He said by lunch time this whole place will be white with snow. He was dead right; by eleven o'clock I was out with the camera, but Dave told me to wait until everything turns white. It was awesome. Dave took me up into the hills in his 4x4 truck and we got some great footage of the snow covered hills.

Treverton School were having a show with Barry Hilton in their school hall and asked if I could accommodate him in the hotel, and can I do the opening act and act as his MC? Our local notice board at our major intersection (?) had '*Our very own Robin Boltman and Barry Hilton; performing.'* I was really adopted in Mooi River.

Before the show a lot of the locals first came into the hotel for a few drinks. There was a girl in the pub I had not seen before. Sparkling eyes first attracted my attention, and a wonderful smile. Her name was Rene. I gave her my card and said, "Mention my name and you'll get a shit seat". She and her family were going to the show and I saw her a few times in the pub after that. Rene and I now live together with her three children, Amy and the twins, Bridget and Claire. I had finally found Love again - or did it find me? Frank Sinatra had a song called, "*Love is lovelier, the second time around.*" Rene was like a soul mate and also my best friend. If you have that in a relationship, where you can share everything, especially your thoughts , everything is just wonderful.

Graham had asked me to do a year at the Argyle, but after my first year, I decided to do another. The Mooi River Tourism Association has a slogan...*come to Mooi River - we will marry you - educate your children - and bury you!* It looks like I'm staying. I've already got a time-sharing bar stool at the famous *Notties Hotel* and of course *the Argyle* is still my local.

One Sunday we went down to Graham Goss, Delyse and Bill near Zulu Falls. Rene and I had a wonderful time with some of the other locals that had also been invited. In the heat of the late afternoon, we were driving along the dust road heading back to *NMR*. (The registration plates of Mooi River) I had a beer going down, but in my rear view mirror, I spotted a local policeman (Mr. Plodd) also heading home from the same party. I put on my hazards and flagged him down. He pulled up behind me and I got out and walked towards his car. *"Problem, Rob?"* I said, *"Yes, I'm drinking and you're not. I don't want you pulling me over for drinking and driving. Have a beer!" "Thanks Rob,"* and we carried on, and had a few more back at the Argyle. I love this place!

Where else can a traffic officer pull you over and say, *"Robin, put on your seat belt!"* Just the other day Rene and I were driving back from Champagne Sports Resort after a show, when this same traffic officer jumped out from behind the bushes near the Hidcote turn-off, just to wave and say hi. Imagine the look of fellow drivers on the N3 - a CY car, waving at a traffic officer, who was waving at us - from behind his special branch… Zeff. He still pulls me over from time to time, *"Put your seat belt on!"*

You could always tell the difference between the cars of sober or drunken people leaving the hotel. The drunks would drive dead straight and the sober ones would weave all over the road, *avoiding the potholes!*

It's a different life on the farm with Rene and the girls, from what I was used to. I even mastered chain saws - we had loads of Black Wattle on the farm. After big cities, cruise liners and travelling the world, doing magic for Ernie Els at Fancourt, Dayle Hayes at San

Lameer, the Million Dollar Golf at Sun City with Pat Symcox - and here I am with fellow farmers burning fire-breaks and chewing on a pigs head at the local! I am slowly settling in - I just need to get some shorts and rugby socks, (just for going out on *formal* nights) then I'll be in the A team!

We still enjoy travelling and every year Allan Foggitt and Starlight Cruises give me Cruise as a present. Graham & I travelled up to surprise Tony de King on his 60th birthday bash and he was led in blindfolded. Loads of tears at that one too, until nearly every entertainer got onto the stage. Some of the Runway Bar locals were also there.

Rene and I went up to Johannesburg one day for Dave Smith-Howells 60th birthday surprise. It was wonderful and loads of entertainers, even Lance James, came out of the woodwork. I had first met Lance in a lift at the Royal Hotel in Durban. We did the *Flood Relief Benefit Concert* in 1988; along with some other entertainers whom you've already met. We stayed with Tony and Nicola, which always has its drawbacks (Just like the Fisher family in Durban)-you never get to bed before three in the morning! We reminisced about how we are at the age now when you get invited to 50th and 60th birthday parties and people ask you to do ten minutes at a funeral!

Durban magic Society, Bernard, Oriole, Dave *Kiwi* O'Conner, Adrian Desmond Smith, Colin & Felicity (Runway bar), Danny and June and the girls, and a host of others have all visited. Joe Parker, Derek Gordon, Lamy and Colin Sumner all came and performed here and the towns folk had a great time on their nights.

Laurie Kay, Wolfgang, Tim and Mariaan, Rowland and Wendy Hobbs, my in-laws, Kevin Hollis (*Sun City),* Alan & Lisa Michie (*Wild Coast),* Rose and Ian, Pete & Noleen, Robbie & Linda (Royal Oak) also came to visit. I have since left the hotel industry to continue with the world of Magic and Entertainment. It's great where we're living now because it is so central to the major areas for shows. The magnificent Drakensberg and the Midlands Meander, where we now live in Rosetta. It gets very cold here in winter and temperatures range from minus 11 degrees Celsius in the morning to a high of 4 degrees by lunch time. It's also close enough to drive to Jo'burg and Durban and I do the odd appearances in Cape Town. It's always good to see the family and friends and have the odd drink at the Royal Oak.

Things don't always go according to plan (as you might have noticed by now). I received a letter from Rene placed under my windscreen *Viper.* This letter was going to bite! I was due to drive down to Balito to do a regular show with Danny Fisher, the *Comedy and Curry* night. Our relationship was over! Maybe it was because my work at her house was complete; carpentry, cupboards in all the rooms and kitchen, the roof, etc.

It's a long drive to the coast and during those two hours I had a lot on my mind. I would have to start all over again, find a place to live, etc. and shortly go on stage and try and be funny and entertaining when your world, as you knew it, had come to an end. The show goes on, as the old adage says, and Danny or the audience were none the wiser. I told Danny afterwards during supper and he was also shocked. His family had been to stay a few times up in the Midlands so often.

I was offered a job at the Barnyard Theatre in Johannesburg with a show called The Rock Circus. It was great and I caught up with a lot of my old friends in JHB. Robbie and Roy Barke from Nedbank days, Rupert *Spook* Hanley and Peter *Bones* Ball etc. And of course Tony and Nicola. Loads of passengers also came to see the show. Rene came up to "properly" say goodbye, but I was homesick, so I left for Cape Town.

ON THE ROAD AGAIN

Back in Cape Town I was offered a job to help my friend Richard Jones with a hotel he was involved in with Queensgate. Up on the west coast is St. Helena Bay and the hotel was a beautiful Hotel, Spa and Golf Estate, Shelley Point. Peter Stuart was the current GM, a wonderful lunatic with a few good people running the place. Martin van der Breggen replaced Peter and together we did our best to run the hotel. My golf improved and I played with a lot of the locals and visiting Royal Oak golfers.

Francois and Franco ran the golf shop and course and Franco had a false leg due to an accident. He was young when it had happened, but it wasn't fitting too well anymore because he was a growing lad. I organised a golf fund raiser and the support was wonderful...we gave Franco a new leg for Christmas.

We visited the famous *Panty Bar* in Paternoster and various other spots along the West Coast. Wonderful down to earth people and I also arranged a few of my entertainer friends to perform there. Rene was still in contact (?) and she came down when Tony and Nicola came to do a show there. They were busy getting their props together in an annex room to the lounge when Tony had placed Nicola's Dolly Parton tits for one of their routines onto a "Shelf". It fell down the laundry chute. He went down to the laundry and asked the wide eyed staff, "Has anybody seen my tits?"

Rupert Mellor also performed on a New Years' Eve party. Martin had left and was working in Lesotho for Sun International now, and was replaced by a guy called Desmond. Queensgate were no longer involved and the place was not the same.

The locals still came around, including Tony Sanderson (Late Nite Live and Chuckle and Chat Show fame)with his wife Jean and the legendary Ian *Mac* MacDonald. Mac is the Papa Surf in the surfing world and started the famous Gunston 500 surfing competition in Durban. Wendy Sharpley (another that got away in my Nedbank days introduced me to Herbal life. I lost about R4000 rand in three days!) Spencer Chaplin was also one of my golfing partners (related to the famous Sir Charlie Chaplin) his sons, he warned me, were also doing wonderful music and would also soon become famous... they did! *Lock'nville.*

Rene was still chatting everyday and wanted me to visit for my upcoming 50th birthday. My birthday party at The Argyle Arms was great and old friends came round - Arnold Geerdts, Dave (Achille) and Craig (Jester), Colin and Felicity (Runway) and Adrian t/a Desmond Durban magician plus the rest! Back at the house it was lovely to see the children again, and the pets. Stitch, the cat is not the friendliest cat, but she jumped off the windowsill onto my lap and just stared into my face. Rene said, "If you don't come back for us, can you at least come back for Stitch?" I went back to Shelley Point and before I could resign I was given a letter of retrenchment from Desmond. I read it in his office and they would be paying me out. I accepted and shook hands and then pulled out my resignation letter and tore it in half and put it on his desk saying, "I guess I wont need this anymore!" The staff threw a farewell party and Tim did another at his home in Cape Town. They all warned me about going back, as did my friends before him on the Achille Lauro, including the captain. Who listens to advice? Only a fool is bitten by the same dog twice, so off I went! The long drive with lots on my mind, back to the Natal Midlands for a second chance.

It was good to be back and I got involved with anything I could do. The kids flourished at school, their dad was now living next door, (my husband in-law) and together we started a business in carpentry, *Botchit and Leggit!* Graham also asked me to work for him again at the Argyle but this time he wanted me to run it and own it. He was getting tired and wanted to sell. A legal contract was drawn up and I would become the owner of the Argyle Arms Country Inn, the business, the building and the land in five years, paying R35 000 a month. I did!

The old staff stayed on and were very happy to have me back, but obviously the manager had to find new employment, which she did and we were due to open soon. Graham closed the hotel to do some renovations but the locals still came in through the back door for drinks. The kitchen took longer than we thought and we delayed the opening by a month. I stayed in the hotel room until Marietta would vacate the managers flat in the hotel.

Opening day was upon us and it was wonderful. Des and Lew Bate did the music and the Karaoke was also a hit. The Midlands people all wished me well with my new venture, but Marietta didn't attend the party. She was packing, etc. for her new job at a wedding venue in the Midlands. The venue phoned me to ask if she had left the hotel yet, because she hadn't pitched up for work. I took the spare key and a female staff member (*Mads)* and went to the flat. After knocking and receiving no response we opened the door and went in. Nothing had been packed and her cell phone and handbag were on her bed. Girls don't leave the house without those two items! I went into the kitchen and looked to the left into the bathroom, and there she was... hanging by a rope around her neck from the shower cross beam! I managed to stop young *Mads* from coming in, led her out and locked the door.

I called the police and Graham and Helen. I called Des to go around to tell Wally (boyfriend) personally. Her Mom came down from neighbouring Escort and police and interviews took forever. Donovan, the owner of the Engen and Wimpy bar in town called me up. (It's a small town and news travels) "Sorry to hear your news, but can you still do the show tonight?" He had asked me to perform for the high school students from all over the province on a cricket tour. He was on the committee and they were all staying in Weston Agricultural School and tonight was their big prize-giving dinner. Rene came to watch over the hotel and the show went on. It's very difficult trying to be funny during a time like this. But I soldiered on and the boys were non the wiser. At the end of the performance Donovan came up on stage to thank me and informed the students about my day. I was given a standing ovation as we walked off together to the Staff Kitchen for a large whiskey!

Back at the hotel quite a few were still left in the bar, and after last rounds Rene spent the night with me. In the words of the Aussie Comedian, Kevin *Bloody* Wilson, "I've had an absolute C*** of a day!"

Rene was a bit of an Angel, Reiki type of a person, and she suggested that she would come around and *Cleanse* my room. She had nightmares for a few days before, saying that, "*She knows what I'm coming to do, and this could get ugly.*" On the day Rene paused before the door and said I must leave, I didn't ask any questions, I was off! Rene did her thing and duly re-appeared, unscathed, and announced that my room had been cleansed and the *Spirit* had left. It was still a week before my two Zulu cleaning ladies would go in. I took two! Rene spent the first night in my new room - I wasn't taking any chances, safety in numbers and all that!

One morning while I was sitting re-wiring my TV and music system, the curtain suddenly blew open and I shot out of there from a seated position!

I got into the business in a big way, and all my friends in the entertainment world would often pop in. Whenever they had gigs in the area they would always do a *turn* at my hotel. The locals loved meeting people they only saw on television, etc. Even The Bats performed at Treverton one evening and they all stayed with me. After their show they announced to the packed auditorium that the party will continue back at "*Rob's Place!"* The pub was really busy that night!

One Sunday a pal, Freddy Loretz drove up from Durban with a few mates. I often stayed at their place as I used to date his wife Fiona's sister, Heather a long time back. Among his pals was the ex Managing Director Alberto Chiaranda from the Wild Coast Sun days. He said he couldn't resist when Freddy said that they were taking a drive up to visit. I showed him around and after the tour, we joined his mates in the bar, and I said, "You always used to sign for your drinks. Do you still have your Magic Pen, Alberto?" He replied with that wry smile and said, "No, but I have a friend who does," pointing at me! I raced upstairs and found my toiletry bag and came back downstairs wearing the Sun International name badge that he had sent round to my dressing room at the Wild Coast, "Mr Magish."

He smiled and said, “You kept that after all these years?”

I don’t think I spent enough time with Rene because I got another letter, just before my birthday, this time on a Facebook message saying it was over, *again!*

I carried on and the hotel flourished. Graham was always paid mostly by halfway through the month. But greed was taking over. The family could now see what their hotel was doing and they became a bit offish!

I took a week off to attend a big school reunion in Cape Town that my old girlfriend from junior school, Milly Ebbing and her friends had organised. All the schools in our area, Junior and High - they were all there, even the shop owner from Roodebloem Road, Bottom Parker! I was their “Guest of Honour” and MC for the night. Students came from all over, USA, England and my sister from Israel. We had a wonderful time at the Upper East Side Hotel.

Back at the Hotel, Graham had taken the decision to buy a few new things that he thought the hotel needed. It wasn’t on my immediate budget plan, but he wanted payment. He also wanted back pay because we started later because of the kitchen floor. I went to the bank, and although I qualified, they warned me. I knew them as friends, but they knew them from business

Friends in the area offered me their lawyer/advocate type of a person who had brought down quite few big rollers in the area. Also a letter was left for me from Graham which I received from a staff member after Graham had left.

She had instructions to hand it to me only after he had left. It was an Ultimatum…pay up or go, according to our "contract"!

I was given until the end of the month, 30th November (that date again) and I weighed my pro's and con's and decided to leave! The locals all knew the whole saga and on my last night we had a good old time and Des announced over the microphone that the Magic is leaving the building! When Graham phoned in the morning to ask about my decision I jangled the key above my phone and said,"Can you hear that? Come and fetch your hotel keys!" He didn't, he sent his daughter Helen and Dave instead!

I was given three hours to vacate the building. We did all the necessary paperwork (all prepared by them) and I cut my losses. I had paid for fourteen months and all the stock too. Unpaid rent from Natalie for the kitchen and restaurant was over R24 000.

I threw in the towel and whatever else I could fit into my little car and left the building, leaving behind loads of my personal effects, like my Achille Lauro life boat oars that Pete Akers had brought up from Cape Town for me. It was the 1st of December, the same date that I left the sinking Achille Lauro! This was going to sink too!

My friend Clive Voss from the Notties Hotel accommodated me and said as soon as his bakkie is back from shopping we'll load all my things and bring it back. I received a cell message from Graham saying that I was banned from coming within 100m of the hotel!

The locals threw a farewell party for me at the Mooi River Country Club and I read out the message from Graham and said, "This applies to all of you!" I had given the Midlands 10 years of my life, it was time to move on and I headed for Cape Town.

I looked forward to being close to the ocean again... it must stem from the "Ancient Mariner" in me. *I must go down to the sea again, the lonely sea and the sky; I left my towel and takkies there, I wonder if they'll be dry?*

A long drive to think about all sorts of things. It was good to be back and on the 12th of December 2012 I stood on the beach at Tableview and waited for the noon gun (all the twelves). On a clear day you can see the puff of smoke on Signal Hill and hear the gun a few seconds later as it drifts over the ocean.

Friends from the Rosetta hotel, my old neighbour from the Midlands were running the St.Helena Bay Hotel and wanted me to come and help. Seeing all those friends again and the crowd from Shelley Point all came around. Business was already improving. Tony (*Late Nite Live* and the *Chuckle & Chat* show) live in the area, along with people like Desiree (Sun City Dancer) and hubby James (the *uncle in the furniture business)* and Ian *Mac* MacDonald, the founder of the famous Durban Surfing Competition, the Gunston 500. (His rep friend did the sponsorship and R500 was the Prize money)

Graham and Mandy Knee from the Wild Coast came to visit and eventually bought a plot and built a beautiful home and B&B called the Bee's Knees and Lazy Daisy's. She is a very talented artist and her paintings are on show in the adjacent studio.

I heard from the friends in the Midlands that Dave Poultney, another friend of Grahams that I had met a few times from Cape Town, had been asked to run the Argyle. Natalie tried for a few months and

Dave took over and lasted two months. He drove to Grahams house and personally handed him the hotel keys. The hotel contents were sent down to Pietermaritzberg by Graham and Helen to go under the hammer at an auction. A few of the locals went down and bought some of my personal belongings and told me they have it in storage for me, including my Achille Lauro life boat oars! Graham died a week later!

My heart was still giving me problems and I moved back to Cape Town after doing a year up on the West Coast. It was time to leave, I thought, when I started dreaming in Afrikaans.

Back home I worked a few shifts behind the bar at the Royal Oak and continued doing Magic. I met some new "locals" and Mike Ferris and his family entered my life. "*Irish Mike"* had another friend from back home and he also became a regular "*Irish Mick"* and his family. To this day I have pleasure in looking after their cars, and its dreadful getting out of theirs and back into my little car!

I was on chronic medication and I had a major water retention problem. I had my dads hands coming out of my sleeves and a Maori rugby players legs coming out of my underpants! Mick asked me to look after his house during a gap with no tenants, just to prevent affirmative shopping. I also had a few chores to do around the house like painting, gardening, etc. The sliding door needed some attention and I discovered the door needed a bolt and a few other things. I went shopping and found a similar one, but I wanted the whole mechanism. I would find that Customer Care lady, they know everything. The bolt kept falling out of my basket while I was shopping, so I popped it into my top pocket. I carried on shopping and R400 later at the till, I was near the door when a guy stopped me and asked me to accompany him to the office. I thought I had won a prize! (customer 1000 for the day or something?) "You have something of ours in your pocket!" I remembered the bolt and said I needed the customer care lady, she'll know where I could get the whole package that should accompany this bolt. But it doesn't work like that. For a bolt that cost's about 45 cents, the police were called and I was taken to the police station. One of the policeman drove my car and I was in the back of the van, following right passed our house! Finger prints and statements and all my dangerous weapons were confiscated, shoelaces, glasses, watch, belt etc. My trousers wouldn't fall down due to my water retention "Maori" legs! I was placed in a holding cell when another staff worker also came in from the same store. He was armed with a KFC box and a coke. He was caught eating in the back of the store. He said the security guy's wife had left him and he was in a "Fowl" mood. I telephoned my good friends Ian and Rose - they would be off today, being a Public Holiday, and they lived nearby. They always seem to be on hand during thick and thin! They took my car home and collected my medication and informed my mom.

Then we were moved to the main cell and I remembered *"Stir Crazy,"* the movie with Gene Wilder and Richard Pryor, "Wha, Wha, I'm bad, I'm bad!" It's not as funny in real life. (*I'll be 126 when I get out*!) The Captain brought my medication and I said that he would have to read the instruction for the dosage. "Don't you know what to take?" I replied, "Yes, but you've taken my glasses!" "Can you show us some magic?" (He had obviously read the statement - job description) I said no because they had confiscated everything out of my pockets, but I borrowed his cigarettes and a coin and did some "miracles" for them. My treatment was a bit better after that, with extra blankets, without too many living organisms!

I had three cell mates; the store guy, one who was curled up and covered and he didn't move the entire night. The talkative one was also called Robin, and he was in for pushing Coke and Heroin and was caught after a car chase. My "bolt" story was fucking pathetic compared to that... a magician that can't even vanish a bolt without being caught!If I had really wanted to steal a bolt, I would make sure it was attached to an Aston Martin! My cell mate was later taken out and released and his lawyer came to fetch him. I would have to spend the night, and maybe the whole weekend, because nothing much gets done on a weekend. It was a Friday, and the Public Holiday was 'Freedom Day!'

Early in the morning, room service came around and gave us coffee. The other guy wasn't dead and we all had coffee together. I kept my plastic spoon - I've watched some movies in my time - I might have to tunnel my way out of here!

Breakfast was served in the en-caged courtyard next to the cell. It's a bit like being inside a big braai with a steel mesh on top. You have no concept of time and ages later Ian and Rose came to visit. I was taken to the glass fronted visitors counter and they wanted to make sure I had survived the night, or even pregnant!

I said that I would probably be in until Monday because it was already so late, but they informed me the detective was here already and it was only 11.00 am, "What time did you guys wake me up? Yesterday?", I shouted to the two *Room Service* policemen on duty. They said to Ian and Rose, "We'll miss this idiot!" I was released with a court hearing date and went home to scrub any living organisms off my body. Back at the Royal I told everybody what had happened. Of course they all thought it was funny as fuck.

Behind the bar it was like, “Can I get you something?” and they would reply, “A beer and a plasma screen!”

I went to court without a lawyer and only spoke once. ”There must be some mistake, Google my name. Have you got the right person or page?” After two appearances the judge informed me that unfortunately they have wasted my time and that everything would be removed from the records and I was free to go. Maybe he did Google me? I said, “Thank you very much. All Rise!” smiled and I did the little bow thing and walked out and the judge just smiled back.

My health was getting into a bad way, and one night I passed out and ended up lying in the bath. My sister Stacy took me to the hospital and Heart Failure and blood pressure problems were diagnosed. After a time in Somerset Hospital, later in the Blaauberg Hospital I was in trouble. I was due to go to The Drakensberg soon for a Maritime Convention, and a book launch about the Oceanos by Andrew Pike titled, “Against All Odds.” The surgeon told me I was not allowed to fly and I won’t be going to the Drakensberg. ”You’ll be going to Groote Schuur Hospital for a Heart Transplant!”

You suddenly go into a totally different mind set. There is so much on in your mind you can’t begin to explain…

Professor Chris Barnard once said to Louis Washkansky’s wife (He was the first recipient in the first Heart Transplant) that the heart is “just a Pump.” Would he still love me after the operation? I love you with all my” heart.” I’m “heart” sore, you’ve broken my “heart,” Shakespeare, ”You have cleft my heart in twain” etc.

How do you say goodbye to something you've had all your life, that's been with you from the beginning through good times and bad? It's been in love more times than I can remember and" broken" so many times, it's skipped a beat on numerous occasions, it's even been up in my throat on sad occasions and pounded like a drum during others. And now we've got to part ways, it's kept me alive for 59 years…

The totally professional team at the Pioneering Groote Schuur Hospital were absolutely amazing. I was in Intensive care for a few weeks. The Nursing staff tolerated my nonsense, magic, covering my face with the sheet as if I had died the night before for the change of shift nurses, wearing my red nose, etc. I would come back after scans or X-rays and announce loudly, "It's a boy, It's a boy!"

Family were absolutely wonderful and came to visit every day. Friends, old and new, army friends and even my army minister from Walvis Bay! It was mentioned on the radio stations and the calls of love, support and concern were truly "heart" warming. I was released because my heart was too weak to operate on just then, and I would have to take the medication and build up a bit of strength by walking and getting the blood moving, etc. I went back once a week and the progress was good; good enough to operate. Instead of a transplant they inserted a pacemaker explaining this is the best first option.

Pete Akers and his team organised a "Have a Heart" Golf day and the response was out of this world. It was scheduled for the 29th November. The operation was done on the 27th Nov. And I was able to be at my Golf Day (Not playing, I was warned) I couldn't anyway.

The fund raising from the friends and the raffle that was organised was overwhelming, to say the least. A friend Maarten also organised a show with Barry Hilton and friends and strangers attended that one too. A girl with photo's of her 21st birthday party, on a cruiser out of Hout Bay with me as her performer; nearly 40 years ago was also present. The guitarist from Jeff Weiner's "Mainstream" band, Mike Larkin, my Achille friend Helen and Cindi (*the one that got away*) were also there.

Allen Foggitt and Daphne donated a cruise on the MSC Orchestra and Richard Jones donated weekend "Getaways" at the beautiful Five Star Simola Golf Resort in Knysna, the St. Helena Hotel plus a host of favourite "eateries and pubs," The Royal, Doodles, Vasco's, Arnold's, Fergie's, Portalia, etc. to name but a few.

Enough money was raised to pay the Medical bills and medications etc. Words are not enough to thank everybody that was part and parcel to these fundraisers. I have also stopped drinking and smoking now, and the worst part of that is not being able to understand any of my friends after 8:00pm!

On my Golf Day I remember during the speeches saying, "It's because of all of you being here today, that I could *also* be here today!"

To quote John Denver…

"Friends, I will remember you, think of you, and pray for you.
And when another day is through, I'll still be friends…with you."

EPILOGUE

The pacemaker is doing fine and during my last check-up, I was told that I have 11 years and 3 weeks battery life. Thank goodness it has an expiry date, I'd hate to be walking around at 163 years old, trying to switch it off… *I won't know anybody!*

It's been great reminiscing with you and sharing my stories, and it's been a privilege to have introduced you to some of the most wonderful people I have met in my travels. When I sit back and think of all we've done, all the fun and laughter we've had and created - and enjoyed every minute of it; and how we had cried, alone or together, on other occasions.

There is talk about making a movie of the Oceanos saga - so if it comes out, please go and see it… *ask to sit in the shallow end!*

Here I am all these years later, after Ian Sinclair had suggested I write this all down one evening at the Argyle Arms. I said I wasn't sure how to go about writing a book, and in the beginning, I told you that I'll just talk and we'll see what happens…

Well, thank you so much for listening - you've been a wonderful audience.

See Ya!

Robin.

PS. *Is a jewel just a pebble that found a way to shine?*

Robin Boltman

Email: robinboltman@gmail.com

Cell: 083 447 7770

Printed in Great Britain
by Amazon

58175772R00115